FUEL ALCOHOL:

How to Make It, How to Use It

Also by the author:

How to Buy a Car

FUEL ALCOHOL:

How to Make It, How to Use It

James R. Ross

ST. MARTIN'S PRESS *New York*

To Barbara Anderson, my editor, for everything she has done to develop and promote my success as an author.

St. Martin's Press, Inc.
175 Fifth Avenue
New York, N.Y. 10010

Library of Congress Cataloging in Publication Data

Ross, James R.
Fuel alcohol, how to make it, how to use it.

1. Alcohol as fuel. I. Title.
TP358.R66 662'.669 80-27793
ISBN 0-312-30932-5
ISBN 0-312-30933-3 (pbk.)

MANUFACTURED IN THE UNITED STATES OF AMERICA

10 9 8 7 6 5 4 3 2 1

Contents

NOTICE

It is illegal to manufacture alcohol without a federal permit (license), and, in some instances, state permit (license) and/or local permit (license). You must secure the proper and necessary permits or licenses prior to manufacturing and using fuel alcohol. (See Chapter Two.)

The alcohol referred to in this book is to be used solely for fuel purposes and is not suitable for drinking. Do not drink it.

It is important that you follow the instructions in this book carefully. Do not begin any procedure described until you are sure you fully understand the instructions and have the proper equipment. It is important that you take seriously all the precautions described in this book. And remember, alcohol is flammable.

We have tried to make this book as accurate and up to date as we could as of the time of publication. However, certain items, such as federal legislation, specific state and local laws and contact information, may change from time to time.

DISCLAIMER

Preface

How would you like to tell the oil companies and utility companies just where to go, and become personally energy independent with fuel alcohol? Sound good? You bet it does. Is it possible? You bet it is. Is it easy? Yes—if you're willing to contribute a little initiative, and a little time, labor, and money.

C_2H_5OH ETHANOL

The ethanol (ethyl alcohol or grain alcohol) revolution is alive and well, and here to stay. It is a viable, renewable-resource energy alternative. It may not be the final answer to *every*one's energy problems, but it is entirely possible for many individuals in the right circumstances to use ethanol for all their energy needs—to generate electricity, heat their homes, and fuel their cars. However, for most do-it-yourselfers, ethanol is only a partial answer; for some, it will mean literally nothing more than being able to choose between regular gasoline and gasohol at the fuel pumps.

Anyone who can boil water and follow simple directions can make alcohol. And anyone who knows the business end of a screwdriver and other simple tools can build the equipment necessary to make alcohol and utilize it for fuel. Unless you live in a population density area, the raw materials necessary to make alcohol are generally readily available throughout the country. So the real question is not can it be done, but rather is it practical for you? This book was written to answer that question and most any other question you might have about fuel alcohol. It was also written to dispel the myriad myths surrounding the current alcohol movement–myths that create both confusion and problems. Fuel alcohol critics have a lot to say against its manufacture and use, some of which is true, most of which is not. As you read this book, you can decide for yourself what is fact, what is fiction, what is good or bad about fuel alcohol, what is right or wrong about it, and how, if at all, fuel alcohol fits into your future.

With so much adverse publicity given fuel alcohol in the late 1970s, it is small wonder that so many people are hesitant to accept it as a viable fuel alternative. It is also quite understandable why a person who has invested $7000 to $10,000 (and more) in a new car or truck is somewhat afraid to take a chance on ruining any part of that vehicle.

Anti-fuel-alcohol zealots create the illusion that if you run your car on fuel alcohol it will begin to corrode and dissolve right before your very eyes. Although alcohol is corrosive (so is water), and is an excellent solvent (so is gasoline), the adverse effects of fuel alcohol on your particular vehicle will not quite happen overnight, if indeed they will happen at all.

To put your mind at ease, consider the following:

The braking system of your vehicle–from the master cylinder to the individual wheel cylinders–contains alcohol (brake fluid).

The cooling system–radiator and internal engine–contains alcohol (antifreeze).

If your transmission leaks, you usually add a can of transmission stop-leak to the transmission fluid without giving it a second thought. Most transmission stop-leak products contain alcohol.

Although these are different types of alcohol than methanol or ethanol, they are, nevertheless, alcohol products that contain carbon, hydrogen, and oxygen with corrosive and solvent properties and characteristics. Eventually, all of the above components–braking, cooling, and transmission systems–will need either maintenance, repair, or replacement, sometimes because of the adverse effects of alcohol, but more often than not because of simple wear and tear caused by friction.

The most common components attacked by fuel alcohol are gaskets, sealants, and diaphragms–either in the fuel pump or carburetor. Eventually, whether you use fuel alcohol or conventional fuel (gasoline or diesel fuel), these components will need maintenance, repair, or replacement. So, the bottom line here is that whether you use alcohol or gasoline, you are eventually going to have problems of one kind or another. In simple economic terms, if you save several hundred or several thousand dollars on fuel costs by using fuel alcohol, is it worth the occasional nominal cost of a few repairs or replacement items on your vehicle—repairs or replacements you would probably have to make anyway if you used gasoline or diesel fuel.

While you hesitate to switch to fuel alcohol, the cost of other fuels—specifically, gasoline, diesel fuel, and home heating fuel—continues to rise. How high must it go before it becomes totally unbearable? It is already totally unreasonable.

Entrepreneurs continue to build multi-million dollar fuel alcohol plants. Some of the investors in these plants are the same people who preach against the individual manufacture of fuel alcohol, claiming it is not feasible both economically and functionally. Who is kidding whom?

Others are currently saving money by making their own fuel, and are becoming, to varying degrees, energy independent. Right now, thousands of farmers are not only growing their own food, but also are making some of their own fuel as well. If it can work for them, it can, in all probability, work for you. But it takes a little time, space, effort, and, of course, money to get started.

A Word of Caution. In the late 1970s, the Department of Energy (DOE) put together some very interesting and sometimes convincing information relative to fuel alcohol production and use. This information is loaded with discrepancies, and is biased in favor of methanol. It is literally impossible for the average person to make methanol, but it *is* possible for the average person to make ethanol. It seems the intent of the DOE is to discourage small-scale individual efforts to obtain energy independence and encourage large-scale corporate control of what will, in the near future, be a major fuel source within this country.

Feasibility studies sponsored by the DOE just don't seem to agree on anything, and feasibility is the one area that will prompt an individual either to pursue fuel alcohol or to simply say, "forget it." Of particular interest is information put out by the DOE on fuel alcohol manufacture energy balance statistics.

Energy balance is a term used to criticize or justify fuel alcohol feasibility. It means that if you put more energy into making alcohol than you get out, you have a negative energy balance and it is not worth the effort of making it. If you get out more than you put in, it is then worth it. The problem with energy balance is that it is usually inaccurately determined. In one report, the DOE uses one method for making alcohol–grain, mashing, and distillation–and uses three different methods for computing energy balance, each of which seems logical and justifiable on the surface. With two of the methods used, DOE derives two negative energy balances, and with the third, it derives a position energy balance. Is the DOE proposing that the individual take the best two out of three, is it merely building a case to discourage small-scale individual pursuit, or is it more accurate to state that the DOE does not really know what it is all about.

Energy balance is generally computed by using Btu (British thermal units) and even here, few agree on the accurate Btu values and ratings involved. One source will give a gallon of ethanol a Btu rating of 72,000, while another will rate it at 84,000 (84,400 Btu is the most commonly accepted). One analyst will rate the fuel energy input for a heat-fired still at 370,000, while another, rating the same operation using the same still, will rate input at 368,000. The more pessimistic ratings are usually given the publicity.

Most consultants, analysts, and upper-echelon critics stop short on energy balance. They generally compute the cost of grain per Btu, and the cost of distillation–fuel cost–per Btu, and give an energy credit for cost recovery of mash (Distiller's Dried Grains–DDG) sold to farmers for livestock feed, but the cow's mouth is where their computations stop. Few follow the feed to the other end of the cow, where an excellent fertilizer is produced that should also be given an energy credit. If not used for fertilizer, cow chips can be dried and used for distillation fuel.

Seldom-considered distillation heat must be given an energy credit if it helps heat a building or shed, as should any heat (which ranges from 80°F to 90°F) generated by the fermentation.

Little consideration is publicly given to using diversified feedstocks for mashing and fermentation, and diversified and combinations of fuel sources, and for using a combination of solar distillation (the fuel is free) and heat-fired distillation to drastically diminish the energy input.

By the time all energy credits are in, a positive energy balance can be obtained regardless of the size of the operation provided there is no intentional waste, and reasonably sound methods are used.

According to a report presented by W. Scheller and B. Mohr to the 171st Annual Meeting of the American Chemical Society, New York, April 7, 1976, it is possible to "achieve a positive energy balance of about 73,000 Btu for each bushel of corn processed into ethanol."

So, regard all negative reports and literature with suspicion and skepticism, look for discrepancies in information, and beware of incomplete information. If something is missing, include it yourself, and draw your own bottom-line conclusions. Accept nothing at face value.

prepared ethanol is currently double the cost of a gallon of gasoline. But that entails construction costs of plant and equipment, overhead and general operating expenses (lights, heat, etc.), salaries, product distribution costs, and, of course, profit. In this respect, the private sector has an advantage over industry. Your place of residence is your place of manufacture, and whether or not you make alcohol, you must have a place to live. You are not particularly involved with salaries and related employee benefits (vacation pay, pension plan, hospitalization, and matching F.I.C.A. taxes). You have no warehousing costs (you can store your product in your basement or in a shed until needed), and no distribution costs. Your construction cost is basically the cost of processing equipment, and the more you are willing to do yourself, the less your total investment will be, which subsequently reduces your total cost per gallon of ethanol produced. Dependent on the cost for raw materials, and the efficiency of your equipment, your cost per gallon of ethanol will come to between 50 and 75 percent the cost of a gallon of gasoline. Much of the percent variance depends on raw-material cost recovery; using the raw material to make alcohol and then selling the residue for animal feed (see cost recovery information, pages 32–34).

HOW TO DETERMINE IF FUEL ALCOHOL PRODUCTION MAKES SENSE FOR *YOU*

Personal fuel alcohol feasibility encompasses many areas: legality, equipment acquisition or construction, raw-material acquisition and processing, disposal of spent materials, and fuel usage, all of which are directly influenced by individual circumstances. Therefore, what is feasible for your neighbor may not be feasible for you.

Legality. As you will see in Chapter Two: How to Get a License, making fuel alcohol for your personal use is absolutely legal. The federal and state red tape is not that great, both in personal time consumption and application procedure. The Bureau of Alcohol, Tobacco, and Firearms (BATF) administers federal-level licensure, and it is with them that you must really start. Most states will not issue a license to make fuel alcohol unless and until you have obtained a federal permit. Your main considerations are: does your background qualify you for a federal permit (no felony convictions, stable background, and adequate references), and do you want to give someone the right to enter your premises (for purposes of search and siezure) without a warrant.

If you have a blemished past, do not let that deter you from considering licensure. A quick call to a regional BATF office with a quick explanation of your particular problem will produce a quick YES or NO direction. Minor traffic-violation convictions have no bearing on your licensability, and anything more serious would be handled on an individual basis. The intent of BATF is not to deny the general public the right to make their own fuel but rather to keep the seedier types from going into a commercial large-scale operation.

The possibility of having your privacy invaded by an unannounced, axe-wielding "revenooer" is extremely slim. Even a casual inspection of your operation is unlikely. The BATF is more concerned with illegal moonshiners, and with legal, large-scale operations rather than with small-scale, personal-use manufacturers. In other words, you are just a drop in the big barrel of alcohol.

Equipment. This could be a major consideration. Mash fermentation tanks (where the yeast and glucose get together to make alcohol) are easily made from easily acquired or salvaged materials. From five-gallon plastic tubs to fifty-five-gallon drums, the cost is negligible. The major equipment cost is in the distillation apparatus: boiler, column, condensing coils, and collection and denaturing equipment. Here you have the choice of buying a ready-made still for several thousand dollars (and up), or making your own for several hundred dollars (and up). If you are a scavenger, mechanically inclined, and innovative, you can develop a "junk-yard," heat-fired still for a very nominal investment and a few days' labor. Consideration may also be given to building a low-cost solar still. Though less expensive, solar stills are also less efficient. The real question is: Do you want to buy or build?

Raw Materials. Where you live has a direct effect on raw-material cost and acquisition. If you are somewhat citified, acquisition presents a minor problem: Will you make enough fuel from X amount of material to justify the time and fuel used to get the material? Even if you live in suburbia, this time/fuel consumption must be considered in your total plan.

The source of your material affects your cost. If you buy from a feed store, obviously the cost will be higher than if you buy direct from a farmer—about 30 percent higher. Ideally, you can work a deal with a farmer to borrow raw material (corn, for example), use it for fermentation, extract your alcohol, and return the fortified residue to the farmer for livestock feed. More commonly called Distiller's Dried Grains (DDG), the residue is 15 to 17 percent richer in food value (primarily protein). Many find it difficult to accept this enriching process, claiming, "you can't take something usable out [alcohol] and leave the base stock [corn] better off for it." While alcohol is a valuable and usable product, yeast, used during fermentation, looks upon it as an undesirable waste product and gladly gives it up, at the same time enriching the base stock by reproduction and the setting aside of by-products. (More on this in Chapter Seven: Mashing.)

Your economic concern is raw-material recovery using DDG, or using the advantages of DDG to convince a farmer to let you borrow his material for a week or so.

Supply-and-demand conditions in the market will always have an effect on the net material cost per gallon, but the total picture should remain proportionately the same.

Processing. Once your equipment is set up, preparation and processing of materials is not that time consuming compared to the benefits of the end product. But it still takes time—several hours per week for the average individual making ten to twenty gallons per week. For the average individual making larger quantities (say fifty gallons per week), the time factor does not increase appreciably, for the basic process remains the same; only the material quantities increase. Nevertheless, this time must be given a value translated into dollars and cents. Again, this is a personal question with a personal answer: How much is your time worth? And is it then more worthwhile to make your own fuel or simply buy it from part of your weekly paycheck?

Location and Space. Where you live and the space you have available may be more important than all of the above considerations. You must have a place for fermentation tanks: Outside, in a shed or open area, or your basement, garage, closet, or an extra bedroom not being used will all do fine. You need cooking facilities for mash preparation and space for measuring and mixing. Most importantly, you must have somewhere to set up a still to distill your alcohol. If you do not have a yard or field, or access to the use of one, fuel alcohol is hardly feasible for you. If you must lease or rent land for distillation, that cost must be added into your total operation on a per gallon basis.

Special Problem. Neighborhoods and neighbors vary greatly. Some are great; everyone gets along well, nobody bothers anyone else, and what you do is entirely your own business. Others are full of people just waiting for the chance to complain or criticize. Regardless of the character of your neighborhood, you must give your neighbors consideration before you establish a still and go into production.

If your town (city or township) has no ordinance regarding fuel alcohol, initiate one. It is possible that you may go into production and subsequently have an envious or spiteful neighbor go to the council and demand that an anti–fuel-alcohol ordinance be passed. You would then lose your investment and chance for energy independence. But it won't happen if you get to the council first. It is more difficult for that envious neighbor to have an ordinance removed from the books than to have one put on. Even though the BATF and state laws may be willing to let you make fuel, a local ordinance against it can shut you down in a minute.

The ordinance need not be lengthy nor written in complicated legalese. Simply put, it can read: "The manufacture of alcohol for fuel purposes shall be permitted within the (city, town, etc.) of (_____) in accordance with state and federal laws governing the manufacture, storage, and use of fuel alcohol; provided that it does not represent a threat to the general public health and safety." Period. That makes it legal and your neighbor can hardly do anything about it. Naturally your neighbor has the right to petition and go for a repeal, but that is hardly likely once the law is on the books.

To approach the council, you can present them with a list of the benefits of fuel alcohol detailed in this book, assure them that your operation will be safe, and will not represent an eyesore to the community (a still is not that beautiful), and that if necessary you will enclose your operation from sight. You might also suggest a ten- to twenty-dollar licensure fee to further induce council to enact the ordinance. Additional income to the city coffers is almost always welcome by the city leaders. It will help if you have a friend or neighbor on your side when you approach the council. And the more the merrier. The council will respond to a group quicker than to an individual.

Your Family. You may be completely sold on the virtues of fuel alcohol; but what about your family? If spouse keeps a spitspot house and immaculately manicured yard, the sudden intrusion of fermentation tanks and a still may create family problems and an unbearably tense situation. It is best to approach this as a family project right from the very start: Sit down with all members present, discuss it, solicit opinions—pros and cons—and clear the air before you go into production. If spouse dissents now, the situation is sure to get worse as you begin to assemble equipment and material. If, however, the entire family pitches in with enthusiasm, your operation will be that much easier. Do not—absolutely not—impose your feelings and opinions on the rest of the family. Just because you think it is a great idea does not mean anyone else does.

Rats. If it has been some time since you have seen (or heard) any furry little animals around your house, get ready for them. Making fuel alcohol will attract rodents—big ones and little ones. When they discover (mostly by smell) what you are doing, they are going to make an appearance sooner or later, and neither you nor your neighbors are going to be too happy about it. So, an ounce of prevention: When you begin your operation, place rat poison strategically around your general work area—grain storage, mashing, and fermenting areas. As they show up, rodents will take the easier-to-get poison and subsequently die off before mating and multiplying. Extreme caution is advised in placing the poison so that it is not accessible to pets and children.

HOW TO DETERMINE IF YOUR CAR AND HOME HEATING SYSTEM CAN BE CONVERTED TO FUEL ALCOHOL USE

It does little good to make alcohol if you cannot use it easily and inexpensively for fuel. Your present car and/or heating system may not be at all compatible to fuel alcohol usage, and may require major changes to make it work—expensive changes that say *forget* fuel alcohol.

Fuel for Your Vehicle. If you own a pre-1975 vehicle, modifying the engine—more specifically the carburetion system—to run on alcohol alone is both simple and practical for the do-it-yourselfer with a modicum of mechanical knowledge. However, modifying the engine is where the simplicity stops and certain intrinsic disadvantages of running a vehicle on pure alcohol begin. Because alcohol has a lower vaporization level than gasoline, cold morning (cold engine) starts present a distinct problem. Engine hesitation and stalling during rapid acceleration occur for the same reason, sometimes even after the engine has reached operating temperature. The corrosion factor on certain engine components–gaskets, diaphragms, fittings, and lines–is extremely high. These are not insurmountable obstacles but rather problems you should consider before making a final decision. A minor problem that could prove major is fuel availability. If you decide to take a long trip, your effective range is limited by the size of your fuel tank and how much spare fuel you can carry in other containers. Fuel alcohol is virtually impossible to find on the open road.

If you own a post-1975 vehicle, you have all of the above problem considerations plus engine modification problems. Thanks to the Environmental Protection Agency (EPA), carburetors on vehicles built since 1974 have become increasingly complex and more difficult to work on or modify. Plugs that inhibit your chances of changing the factory-set lean fuel-air mixture are permaset over and against the fuel-air mixture screws on some carburetors (fuel-air mix and idle screw must be adjusted for alcohol conversion). Pollution devices also complicate the conversion process, as do the carburetor altitude compensators found on some models made between 1975 and 1977. There are other considerations, which will be covered in more detail in Chapter Eleven: How to Convert Your Car (or Truck) to Run on Total Alcohol.

Remember: conversion of a post-1975 car to alcohol usage is not impossible–just more difficult than conversion of a pre-1975 car.

Gasohol. Perhaps the easiest and most desirable use of alcohol to fuel a vehicle is in gasohol–more as a gasoline extender than a total fuel replacement. Gasohol is a mixture of alcohol and unleaded gasoline, the most common mixture being 10 percent alcohol and 90 percent gasoline. However, mixes of up to 30 percent alcohol and 70 percent gasoline can be used successfully with no engine (carburetion) modifications. The only change you may need to make is in the fuel filter. Alcohol is an excellent solvent and will dissolve any varnish and/or gum buildup caused by gasoline between the fuel tank and fuel filter. Some of these dissolved residues can lodge in the filter and restrict the flow of fuel to the carburetor. The engine may then stall, hesitate, or kick for lack of enough fuel. However, after a few fill-ups with gasohol, the problem is generally nonrecurring.

Late-model cars with small fuel filters in the carburetors are more susceptible to this clogging than those with larger in-line filters. In any event, if this problem is going to develop, it will eventually occur whether you use a 10-percent or a 30-percent blend. Naturally, if you go back to regular gasoline for a prolonged period of time and then return to gasohol, the filter may again clog and need changing. But this is a minor consideration–easy to do and inexpensive.

Alcohol, working as a solvent, will keep most components past the fuel filter clean. Once past the carburetor, alcohol will burn cleaner than straight gasoline, produce less carbon inside the engine, and generate fewer exhaust pollutants (the primary pollutant alcohol generates is aldehyde, which is generally taken care of by the catalytic converter portion of your exhaust system). Making your own gasohol is covered in Chapter Ten: Using What You Have Made to Fuel Your Car or Truck.

By now you should be ready to take a quick look at the advantages and disadvantages of fuel alcohol to best determine your requirements compatibility.

Fuel for Your Home Heating System. The bottom line on heating-system fuel alcohol usage comes down to a go/no-go decision. The following considerations must be made before you even begin to pursue an alcohol-fired heating system:

1. Do you currently have a liquid-fuel-fired heating system compatible to conversion?
2. Do you have a hot water heating system that can be easily converted to liquid fuel?
3. Gas-, coal-, or wood-fired hot water heating systems are compatible to fuel alcohol conversion at nominal cost and effort.
4. Electric heating systems require the installation of a totally different heating system at considerable cost and effort (labor).
5. What is the pay-back period in years for alcohol conversion? In other words, when will the money you save by going to fuel alcohol balance the investment made (input) for conversion?
6. Will you live in the house long enough to recapture the investment? See Chapter Twelve: Using What You Have Made to Fuel Your Home Heating System, for more details.

HOW TO DETERMINE HOW MUCH FUEL ALCOHOL YOU NEED TO PRODUCE

Begin with your current needs—how much gasoline and/or home heating fuel do you now use? That amount is then gauged against the respective heat values of alcohol and gasoline or fuel oil. To replace gasoline with alcohol, you will use 10 to 15 percent more alcohol than gasoline to drive the same number of miles, since the Btu value of alcohol is about 84,000, and of gasoline about 117,000 to 124,000. Arithmetically, the Btu of gasoline is 29 to 33 percent higher than that of alcohol. However, alcohol burns leaner in an internal combustion engine (your car or truck) because of its higher oxygen content. Alcohol also has a higher octane rating (98 to 100) than gasoline (86 to 91). So, if you currently use 20 gallons of gasoline each week, you will need 22 to 23 gallons of alcohol to replace it.

If you plan to make gasohol, you will need to make only two to six gallons of alcohol per week depending on the percent gasohol blend you plan to make. (See Chapter Ten: Using What You Have Made to Fuel Your Car or Truck.)

To replace home heating fuel oil with alcohol, you will need about 20 to 25 percent more alcohol by volume, since fuel oil has a Btu rating of 140,000. If you currently use 200 gallons of fuel oil per month, you will need to make 240 to 250 gallons of alcohol. Of course you may decide to add a percentage of alcohol to your fuel oil using alcohol more as a fuel extender than a fuel replacement. (See Chapter Twelve: Using What You Have Made to Fuel Your Home Heating System.)

HOW TO DETERMINE HOW MUCH SPACE YOU NEED FOR EQUIPMENT

Each gallon of alcohol you need requires a certain amount of space for processing, fermentation tank, distillation equipment, etc. The two major variables of equipment space are how many times per week the equipment is used, and equipment distribution. If you use the equipment three times per week, you can use equipment designed to make 10 gallons of alcohol to make 30 gallons of alcohol. Multiple weekly runs require a little more labor and time, but save a lot on space and equipment size needs. If you distribute your equipment, i.e., you have your fermentation tank in your basement, your processing equipment in the kitchen, your distillation equipment in the back yard, and your product storage in the garage, you can get by with minimal space. Equipment distribution is ideal if you are on a space budget. Even so, alcohol production still requires space. So how much space do you need?

Using 20 gallons of alcohol per week as a base, you will need a 4′ × 4′ area for processing; a 4′ × 4′ area for a fermentation tank; a 4′ × 10′ area for distillation; and an additional 4′ × 11′ area for sprouting, material storage, and a workbench. So, for a minimum of 20 gallons of alcohol per week, you should plan on a total space requirement of 12′ × 22′. If this is squeezed into a shed according to the floor plan shown in Figure 1A, you will need at least an eight-foot ceiling clearance for the distillation equipment.

If you want to double your production capacity to 40 gallons per week, the external linear measurements of your equipment do not double. To make 20 gallons of alcohol per week, you need a fermentation tank 4′ × 4′ × 4′ (a four feet cube), which is 64 cubic feet. For 40 gallons per week, you need add only one foot in all directions (a five feet cube), which is 125 cubic feet. Need another 20 gallons? Add another foot (a six feet cube) for 216 cubic feet—about 60 gallons per week.

MAJOR DISADVANTAGES

1. **Corrosiveness.** Both methanol and ethanol are corrosive. However, methanol is the more corrosive since it contains about 50 percent oxygen (which increases the rate of oxidation, i.e., corrosiveness); while ethanol is about 33 percent oxygen. Both act as solvents on polyurethane, certain plastics, gaskets, metals, etc.
2. **Automotive performance.** The higher the percent of alcohol, used either as a blend or as pure fuel, the more problems realized by the user. Corrective measures must be taken to overcome cold morning starts, engine hesitation, and stalling.
3. **Fuel separation.** Water, usually drawn from the atmosphere, will have a tendency to separate the alcohol from the gasoline because the bond between alcohol and water is stronger than the bond between alcohol and gasoline. This is more a commercial concern where large quantities of gasohol are stored for relatively long periods of time, or where the sheer volume of storage tanks induces water absorption. Proper mixing, venting, and storing reduces the magnitude of the problem. For the

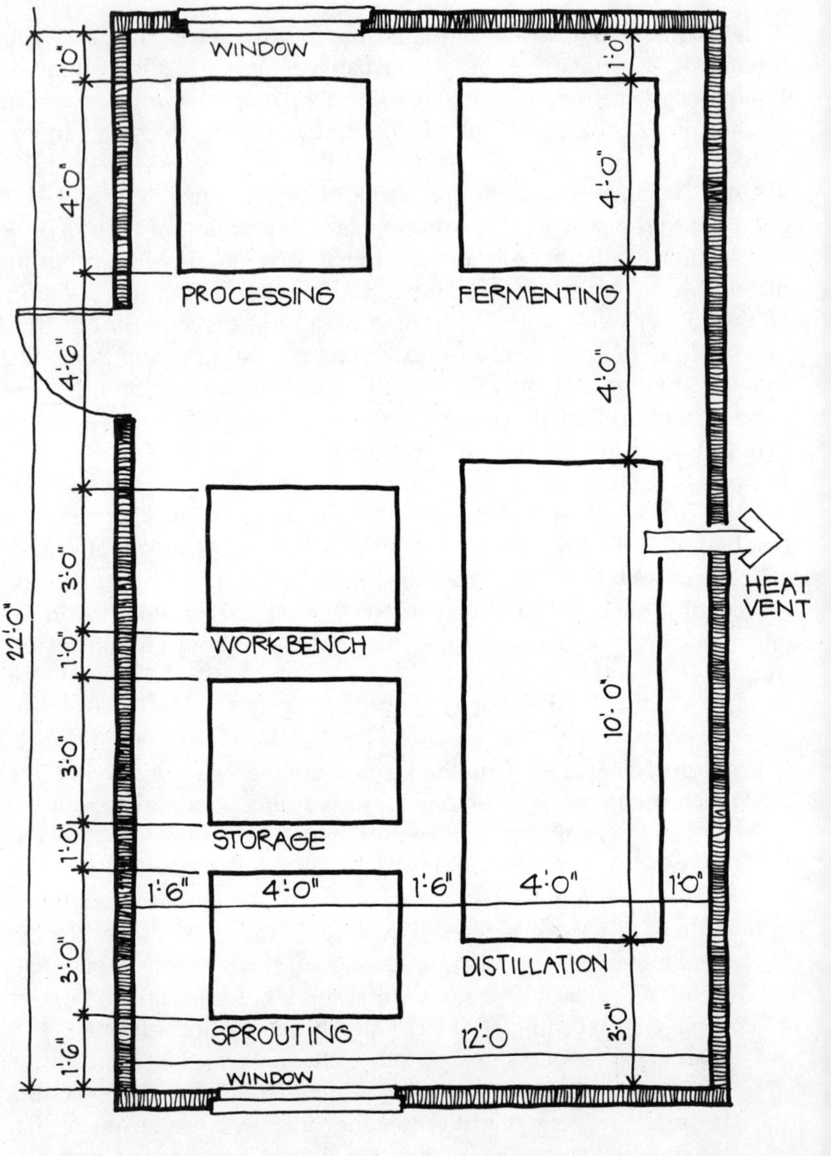

Figure 1A. Basic Alcohol Production Floor Plan

small-scale operation, fuel separation does not pose much of a problem.

4. **Vapor pressure.** Straight gasoline evaporates and expands when stored, creating vapor pressure that must be properly vented. Gasohol creates a higher vapor pressure than straight gasoline—a pressure which may exceed certain state laws. This is again primarily a large-volume commercial-operation problem. Special venting must be used which both vents fumes and pressure and excludes atmospheric moisture. Your small-scale operation storage may be affected if your state places alcohol in the same category as gasoline—Class I, Flammable Liquid—as South Dakota does. It is best to check it out (see Chapter Two: How to Get a License, and Contact Information).

MAJOR ADVANTAGES

1. **Renewable resource.** Ethanol can be produced from one or a combination of several of the many crops grown within this country, without lowering food availability to the populace. When farm subsidies are dropped, more crops will be grown to accommodate both the ethanol and food needs of the country. Naturally, crops can be personally grown and used on an individual basis, obviating dependency on an external source. When food production drops, cellulose (forestland) is available for ethanol production.
2. **Farm surplus and spoilage.** Ethanol production offers the farmer an outlet for any surplus he may have because of market saturation. And the farmer need no longer worry about crop spoilage—that too can be fermented into ethanol.
3. **Livestock feed.** Ethanol produced by grain fermentation yields a by-product high in protein and suitable for livestock feed (about 15 percent to 17 percent higher feed value than the original grain).
4. **Other by-products.** Ethanol made by fruit or vegetable fermentation yields a by-product suitable for compost and fertilizer. In certain instances it can be used for livestock feed—pigs will eat most anything. Bagasse, a left-over by-product of sugar cane when the juice has been extracted, can be used for fuel, or compost, or to make fiberboard, all of which represent a cost recovery credit.

5. **Storage.** Alcohol can be stored longer and less dangerously than other fuels–specifically, gasoline. While gasohol has a higher vapor pressure than gasoline, alcohol has very little vapor pressure (look at the bottle of alcohol stored on your medicine cabinet shelf). Under certain conditions during usage, gasohol can cause vapor lock (see Chapter Ten: Using What You Have Made to Fuel Your Car).
6. **Production.** Almost anyone can make ethanol. When is the last time you tried to make gasoline, diesel fuel, or home heating fuel oil?
7. **Efficiency.** Alcohol has a lower Btu rating than fuels it will replace; but, when blended with gasoline, it acts as an octane enhancer. For each three percent of alcohol blended with gasoline the octane rating goes up about one point. This means you can buy cheap, low-grade (low octane) gasoline and blend in alcohol to make a high-grade fuel. Performance factors and fuel economy will vary depending on the type vehicle and the state of tune of the engine. Variance, plus or minus, is not really that significant–about five percent.
8. **Pollution.** Used in an internal combustion engine, gasohol and alcohol are less polluting–less of a threat to the environment–than gasoline and diesel fuel. Used as home heating fuel in place of fuel oil, they are again less polluting. However, when used to replace natural gas or electric heat, they are more polluting.
9. **Independence.** During times when existing oil supplies are dwindling, alcohol offers a viable source of fuel and energy–an alternate that is not externally controlled. With rapidly increasing oil prices, alcohol offers economy and savings, and both personal and national energy independence.

PERSONAL DETERMINATION CHECKLIST

		YES	NO
1. Legality:	Do you qualify for a license?	_____	_____
2. Equipment:	Are you prepared to build or buy?	_____	_____
	Does the cost justify the product?	_____	_____
3. Raw Materials:	Do you have easy access?	_____	_____
	Is the cost reasonable?	_____	_____
	Is cost recovery possible (DDG outlet)?	_____	_____

	YES	NO
4. Processing: Do you have the time?	_____	_____
Are you willing to take the time?	_____	_____
5. Location: Do you have space for all equipment?	_____	_____
6. Is Production compatible with your neighborhood?	_____	_____
7. Is your family in favor of alcohol production?	_____	_____
8. Can you easily use the fuel you produce?	_____	_____

If you have more Noes than Yeses, fuel alcohol is probably not for you, particularly if you answered No to either question one or five. Any other No answers (obstacles) can generally be handled with a little ingenuity and initiative.

USAGE DETERMINATION

	YES	NO
1. Do you have the mechanical abilities to make engine modifications?	_____	_____
2. Are you willing to pay a mechanic to make engine modifications?	_____	_____
3. Do you have a pre-1975 vehicle?	_____	_____
4. Do you have an oil-fired hot water heating system?	_____	_____
5. Do you have a hot water heating system fired by any other type fuel?	_____	_____

Noes to questions one, two, and three fairly well eliminate you from running your car on 100-percent alcohol. Noes to questions four and five pretty much eliminate use of home-heating fuel alcohol. That leaves making gasohol as your primary (if not only) use of alcohol. Your main question is then reduced to: Is it feasible and practical for you to make your own alcohol to extend your motor fuel 10 to 30 percent, or, in other words, is it worth the trouble to you to make alcohol and save 5 to 15 percent (10- to 30-percent gasohol, respectively) on your motor fuel?

2

How to Get a License—and Other Legal Considerations

IT'S THE LAW

YOU MUST HAVE a license to produce alcohol, even if it is for your own personal use. If you are the type who hates to fill out credit card applications, and despises even the short form for filing your income tax, legislative requirements for licensure may deter you from pursuing fuel alcohol production. Applying for a license is not really that confusing, nor is it overcomplicated with red tape. The process is basically simple for an experimental permit, which is what you need for personal-use alcohol production. However, obtaining a distilled-spirits plant permit (commercial permit), which authorizes alcohol production and sale/distribution, is a bit more complex.

Experimental Permit. An application for this permit must be filed with the Bureau of Alcohol, Tobacco, and Firearms (BATF) regional office (see directory at the end of this chapter). This permit—once granted—authorizes you to produce alcohol for your personal use only. Generally, even if your state has a fuel alcohol licensing law, once the federal permit is issued, the rest is easy—your state will issue a state license. If your state has no license law (see table, page 28), you can begin production as soon as BATF says, "GO." In any event, having a federal permit does not exempt you from state and local regulations, particularly on matters of air and water pollution, noise pollution, safety factors, and general public welfare.

How to Apply. Initially, a simple telephone call to your regional BATF office requesting an application form for an experimental permit will suffice to get you started. You will receive the form shown on pages 17–20.

DEPARTMENT OF THE TREASURY
BUREAU OF ALCOHOL, TOBACCO AND FIREARMS

APPLICATION AND PERMIT FOR ALCOHOL FUEL PRODUCER

(Prepare in Duplicate, See Instructions)

ATF USE ONLY

DATE RECEIVED	DATE RETURNED AFTER CORRECTIONS
PERMIT NUMBER	EFFECTIVE DATE

1. ORIGINAL APPLICATION *(Check applicable box)*

☐ SMALL – 10,000 Proof Gallons or less*

☐ MEDIUM – More than 10,000 Proof Gallons but not more than 500,000*

☐ LARGE – More than 500,000 Proof Gallons*

*Proof Gallons to be produced and received during one calender year *(See Instruction 4.)*

2. AMENDED PERMIT *(Check applicable box(es))*

(Change In)

☐ LEVEL OF OPERATIONS

☐ OFFICERS OF CORPORATION *(Medium and Large Plants Only)*

☐ CONTROL OF CORPORATION *(Medium and Large Plants Only)*

(Change In)

☐ NAME OF PROPRIETOR

☐ LOCATION OF PLANT

☐ OTHER *(Explain)*

PERMIT NO

AFP–

3. NAME OF OWNER *(If partnership, include name of each partner)*

4. DAYTIME TELEPHONE NUMBER *(Include Area Code)*

5. LOCATION *(If no street address show rural route)*

6. MAILING ADDRESS *(If different from plant location) (RFD or Street No., City, State, ZIP Code)*

7. PREMISES FOR ALCOHOL FUEL PLANT ARE *(Check applicable box)*

☐ OWNED BY THE APPLICANT *(Skip Item 8.)*

☐ NOT OWNED BY THE APPLICANT *(Complete Item 8.)*

8. OFFICERS OF THE BUREAU OF ALCOHOL TOBACCO AND FIREARMS ARE GRANTED ACCESS TO THE PREMISES DESCRIBED BY THIS APPLICATION FOR A ALCOHOL FUEL PRODUCERS PERMIT.

NAME OF PROPERTY OWNER

SIGNATURE OF/FOR PROPERTY OWNER	DATE

9. STILLS FOR FUEL PRODUCTION ON THE PLANT PREMISES

STILL MANUFACTURER *(If owner is the manufacturer, write "Owner")* (a)	SERIAL NUMBER OF STILL (b)	KIND OF STILL *(State whether Kettle, Pot, Charge Chamber, Continuous Still, or Other (Specify))* (c)	CAPACITY *(Proof Gallons) (See Instruction 7.)* (d)

10. BASIC MATERIALS *(Other than yeasts or enzymes)* TO BE USED IN PRODUCTION OF SPIRITS *(Check applicable box(es))*

☐ GRAIN *(Corn, Wheat, Sorghum, Barley, etc.)* AND OTHER STARCH PRODUCTS *(Potatoes, Sweet Potatoes, etc.)*

☐ SUGAR BASED CROPS *(Cane Sugar, Sugar Beets, Molasses, Sweet Sorghum, Beet Fodder, etc.)* AND PRODUCTS

☐ FRUITS OR FRUIT PRODUCTS *(Grapes, Peaches, Apples, etc.)*

☐ FORAGE CROPS *(Alfalfa, Sudan Grass, Forage Sorghum, etc.)*

☐ CROP RESIDUE *(Garbage or other refuse)*

☐ OTHER *(Specify)* ____________________

11. DESCRIBE SECURITY MEASURES *(Such as use of locks, fences, building alarms, etc.)* TO PROTECT PREMISES, CONTAINERS, STILL(S) AND BUILDINGS WHERE SPIRITS ARE STORED

12. DIAGRAM OF PLANT PREMISES *(In the space provided or by attached map or diagram, show the area to be included for the alcohol fuel plant. Identify roads, streams, lakes, railroads, buildings, and other structures or topographical features on the diagram. Show location(s) where alcohol fuel plant operations will occur. The diagram should be in sufficient detail to locate your operations and premises.) (See directions for sample diagram.)*

13. I WILL COMPLY WITH THE CLEAN WATER ACT *(33 U.S.C. 1341(a)). (Will not discharge into navigable waters of the U.S.)*

☐ YES ☐ NO

14. IF THIS APPLICATION IS APPROVED AND THE PERMIT IS ISSUED, I CONSENT TO THE DISCLOSURE OF THE NAME AND ADDRESS SHOWN ON THE APPLICATION IN AN ATF PUBLICATION, "ALCOHOL FUEL PRODUCERS", WHICH MAY BE DISTRIBUTED ON REQUEST TO THE GENERAL PUBLIC *(including media, business, civic, government agencies, and others)*. UNDER 26 U.S.C. 6103 YOU HAVE A LEGAL RIGHT NOT TO GIVE THIS RELEASE.

☐ YES ☐ NO *(A no response will have no effect on the consideration given this application)*

APPLICANTS FOR MEDIUM AND LARGE ALCOHOL FUEL PLANTS MUST ATTACH THE ADDITIONAL INFORMATION REQUIRED IN INSTRUCTIONS

Under the penalties of perjury, I declare that I have examined this application, including the documents submitted in support thereof or incorporated therein by reference, and, to the best of my knowledge and belief, it is true, correct, and complete.

15. SIGNATURE OF/FOR APPLICANT	16. TITLE *(Owner, Partner, Corporate Officer)*

STOP
MAKE NO FURTHER ENTRIES ON THIS FORM

ALCOHOL FUEL PRODUCERS PERMIT
UNDER 26 U.S.C. 5181

1. EFFECTIVE DATE

2. PERMIT NUMBER

AFP –

Pursuant to the above application and subject to applicable law and regulations and to the conditions set forth below you are hereby authorized and permitted at the premises described in your application to produce, process and store, and use or distribute distilled spirits (Not including distilled spirits produced from petroleum, natural gas or coal) exclusively for fuel use. The quantity to be produced and received from other plants during the calendar year is limited to the quantity stated in this application.

This permit is continuing, and will remain in force until suspended, revoked, voluntarily surrendered, or automatically terminated. This permit does not allow you to operate in violation of State or local laws.

THIS PERMIT IS NOT TRANSFERABLE. In the event of any lease, sale, or other transfer of the operations authorized, or of any other change in the ownership or control of such operations, this permit shall automatically terminate. (See 27 CFR 19.145 AND 19.949)

3. SIGNATURE OF REGIONAL REGULATORY ADMINISTRATOR, BUREAU OF ALCOHOL, TOBACCO AND FIREARMS

CONDITIONS

1. That the permittee in good faith complies with the provisions of Chapter 51 of Title 26 of the United States Code and regulations issued thereunder.

2. That the permittee has made no false statement as to any material fact in his application for this permit.

3. That the permittee discloses all the material information required by law and regulation.

4. That the permittee shall not violate or conspire to violate any law of the United States relating to intoxicating liquor and shall not be convicted of any offense under the United States Code punishable as a felony or of any conspiracy to commit such an offense.

5. That all persons employed by the permittee in good faith observe and conform to all of the terms and conditions of this permit.

6. That the permittee engages in the operations authorized by this permit within a 2 year period.

7. This permit is conditioned on compliance by you with the Clean Water Act (33 U.S.C. 1341(a)).

DEPARTMENT OF THE TREASURY – BUREAU OF ALCOHOL, TOBACCO AND FIREARMS

APPLICATION AND PERMIT FOR ALCOHOL FUEL PRODUCER UNDER 26 U.S.C. 5181

INSTRUCTION SHEET FOR ATF FORM 5110.74

PLEASE READ CAREFULLY. AN INCOMPLETE OR INCORRECT APPLICATION WILL DELAY YOUR ALCOHOL FUEL PRODUCER'S PERMIT.

1. PURPOSE. The application is completed by a person (applicant) who would like to establish a plant to produce, process, and store, and use or distribute distilled spirits to be used exclusively for fuel use under 26 U.S.C. 5181. Distilled spirits means only ethanol or ethyl alcohol. The production of methanol does not require a permit from the Bureau of Alcohol, Tobacco and Firearms. The production of distilled spirits from petroleum, natural gas, or coal is not allowed by the Alcohol Fuel Producers Permit.

2. GENERAL PREPARATION. Prepare this form and any attachments in duplicate. Use separate sheets of approximately the same size as this form when necessary or as required. Identify these separate sheets with your name, and attach to form.

3. WHERE TO FILE. Submit application to the appropriate regional regulatory administrator. (A listing of their offices and the areas they are responsible for is on the reverse.)

4. TYPE OF APPLICATION. (See item 1. of application). The type of permit you need, small, medium, or large, depends on how many proof gallons of distilled spirits you intend to produce and receive by transfers from other plants during one calendar year. The number of proof gallons may be calculated by taking the proof of the spirits multiplied by the wine gallons (your standard American gallon) and dividing by 100. For example:

   50 gallons of 190° proof spirits =
   190 x 50 ÷ 100 = 95 proof gallons.

   After determining the quantity of spirits to be produced and received by transfers from other plants, check applicable box in item 1.

5. AMENDED PERMIT. Item 2. is only completed when changing or amending an existing permit.

6. INFORMATION ABOUT APPLICANT CURRENTLY ON FILE WITH REGIONAL REGULATORY ADMINISTRATOR NEED NOT BE RESUBMITTED. State in item requesting such information the type and the number of the license or permit for which the information was filed.

7. CAPACITY OF STILLS. Item 9.d. asks for the capacity of your still(s) in proof gallons. The capacity should be given as the greatest number of proof gallons of spirits that could be distilled in a 24 hour period. Capacity of column still may be shown by giving the diameter of the base and the number of plates or packing material. The capacity of a pot or kettle still may be shown by giving the volumetric (wine gallon) capacity of the pot or kettle.

8. SAMPLE DIAGRAM OF PREMISES. Item 12. requires a diagram of your plant premises. It can be drawn by hand and does not have to the drawn to scale. Below is a sample of such a diagram.

9. ITEM 15—SIGNATURE OF/FOR APPLICANT. (a) Individual owners sign for themselves; (b) partnerships have all partners sign, or have one partner who has attached an authorization to act on behalf of all the partners (unless this authorization is provided by State law); (c) corporations have an officer, director, or other person who is specifically authorized by the corporate documents sign; (d) any other person who signs on behalf of the applicant must submit ATF F 1534 (5000.8), Power of Attorney or other evidence of their authority.

10. ADDITIONAL INSTRUCTIONS FOR APPLICANT OF SMALL ALCOHOL FUEL PLANT. Complete all items on the application form. Be sure that you sign and date the form in items 15, and 16. NO ADDITIONAL INFORMATION IS REQUIRED.

11. APPLICANT FOR PERMIT OF MEDIUM AND LARGE ALCOHOL FUEL PLANT. Complete all items on the form. Supply in duplicate the additional information and forms as stated below.
    a. The following information is requested for an individual proprietor, each partner, or each director of a corporation or similar entity and each officer of the corporation or similar entity who will have responsibilities and connection with the operations covered by the permit. In addition applicants for large plants shall show the same information for each interested person who is an individual as listed in the statement of interest in the organizational documents required by 27 CFR 19.941.
       (1) full name including middle name;
       (2) title in connection with applicants' business;
       (3) social security number;
       (4) date of birth;
       (5) place of birth; and
       (6) address of residence.
    b. A statement as to whether the applicant or any person required to be listed by the instructions above has been previously convicted of any violation of Federal or State law (other than minor traffic violations).
    c. A statement of the maximum quantity of distilled spirits to be produced and received from other plants during a calendar year.
    d. A Distilled Spirits Bond, ATF F 5110.56, as required by 27 CFR 19.956.
    e. Any other information required by the regional regulatory administrator after examination of this application.

SPECIAL INSTRUCTIONS FOR ALL APPLICANTS

12. OPERATIONS BEFORE ISSUANCE OF PERMIT. Unless otherwise specifically authorized by law or regulations, an applicant for an Alcohol Fuel Producer Permit may not engage in operations until the permit has been issued by the regional regulatory administrator.

13. STATE AND LOCAL LAWS. This permit does not allow you to operate in violation of State or local laws. Applicants should check with appropriate State and local authorities before engaging in alcohol fuel plant operations.

14. ATF FORMS AND REGULATIONS FOR ALCOHOL FUEL PLANTS. ATF forms and publications including regulations for alcohol fuel producers may be ordered from the ATF Distribution Center, 3800 S. Four Mile Run Drive, Arlington, VA 22206.

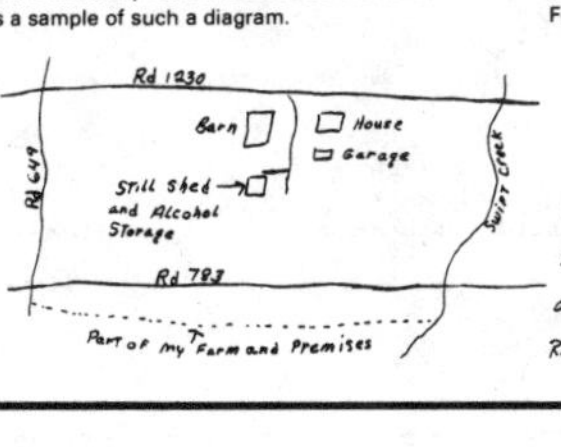

My farm is 1,200 acres, 1,000 acres are bounded by Rd. 649, 1230, 783 and Swift Creek. 200 acres are just the other side of Rd 783.

FOR QUESTIONS CONCERNING YOUR APPLICATION CONTACT YOUR REGIONAL REGULATORY ADMINISTRATOR AT THE ADDRESS SHOWN BELOW.

IF APPLICANT'S BUSINESS IS TO BE CONDUCTED IN:	CALL OR WRITE: REGIONAL REGULATORY ADMINISTRATOR ATTN.: CHIEF, TECHNICAL SERVICES BUREAU OF ALCOHOL, TOBACCO & FIREARMS
Connecticut, Maine, Massachusetts New Hampshire, New York, Rhode Island, Vermont, Virgin Islands, Puerto Rico	6 World Trade Center, 6th Floor New York, New York 10048 PHONE: NY ONLY 800-442-8275 ALL OTHER 800-223-2162
Delaware, District of Columbia, Maryland, New Jersey, Pennsylvania, Virginia	2 Penn Center Plaza Philadelphia, Pennsylvania 19102 PHONE: PA ONLY 800-462-0434 ALL OTHER 800-523-0677
Alabama, Florida, Georgia, Mississippi, North Carolina, South Carolina, Tennessee	P.O. Box 2994 Atlanta, Georgia 30370 PHONE: GA ONLY 800-282-8878 ALL OTHER 800-241-3701
Indiana, Kentucky, Michigan, Ohio, West Virginia	550 Main Street Cincinnati, Ohio 45202 PHONE: OH ONLY 800-582-1880 ALL OTHER 800-543-1932
Illinois, Iowa, Kansas, Minnesota, Missouri, Nebraska, North Dakota, South Dakota, Wisconsin	230 S. Dearborn Street Chicago, Illinois 60604 PHONE: IL ONLY 800-572-3178 ALL OTHER 800-621-3211
Arkansas, Colorado, Louisiana, New Mexico, Oklahoma, Texas, Wyoming	Main Tower 1200 Main Street Dallas, Texas 75202 PHONE: TX ONLY 800-442-7251 ALL OTHER 800-527-9380
Alaska, Arizona, California, Hawaii, Idaho, Montana, Nevada, Oregon, Utah, Washington, Guam	525 Market Street San Francisco, California 94105 PHONE: CA ONLY 800-792-9811 ALL OTHER 800-227-3072

PRIVACY ACT INFORMATION

The following information is provided pursuant to Section 3 of the Privacy Act of 1974 (5 U.S.C. 552(a)(e)(3)):

1. AUTHORITY. Solicitation of this information is made pursuant to 26 U.S.C. 5181. Disclosure of this information by the applicant is mandatory if the applicant wishes to obtain an Alcohol Fuel Producers Permit.

2. PURPOSE. To determine the eligibility of the applicant to engage in permit operations, to determine location and extent of the permises, and to determine whether the operations will be in comformity with law and regulations.

3. ROUTINE USES. The information will be used by ATF to make determinations set forth in paragraph 2. In addition, the information may be disclosed to other Federal, State, foreign and local law enforcement and regulatory agency personnel to verify information on the application where such disclosure is not prohibited by law. The information may further be disclosed to the Justice Department if it appears that the furnishing of false information may constitute a violation of Federal law. Finally, the information may be disclosed to members of the public in order to verify the information on the application where such disclosure is not prohibited by law.

4. EFFECTS OF NOT SUPPLYING REQUESTED INFORMATION. Failure to supply complete information will delay processing and may result in a denial of the application.

The following information is provided pursuant to Section 7(b) of the Privacy Act of 1974:

Disclosure of the individual's social security number is voluntary. Pursuant to the statutes above, ATF is authorized to solicit this information. The number may be used to verify the individual's identity.

Once the regional BATF office has reviewed your application for completeness, it is forwarded to a field officer for subsequent field inspection. The probability of a field inspection on a small-scale operation is very remote. Inspections are primarily on those operations requiring a bond and on commercial operations. If you are subject to inspection, do not try to butter up or otherwise influence the inspector; and absolutely do not try to bribe.

It is advantageous to secure or build your equipment–distillation, fermentation, collection and storage apparatus–prior to inspection, for it will speed processing and give the inspector a chance to offer constructive advice or criticism. (Details on construction, size, and capacities, will be given in later chapters.) (NOTE: In California, you must have a license to own a still or distillation apparatus, whether or not it is being used, so you cannot secure or build a still as suggested above. Your first step in California is to secure a California Still License, and go from there.) To be on the safe side, check your state requirements regarding still ownership. In California it is a felony to own a still without a license.

Consideration must be given to the fact that you may build or secure your equipment and subsequently be denied a license, either state or federal. Do not waste the time and money on equipment if there is any doubt whatsoever about your ability to secure the proper license(s).

The regional BATF office will issue a formal authorization (permit) upon receipt of: a favorable inspection report (if an inspection is indeed performed); a favorable report based on a cursory background check; and, if applicable, a properly fulfilled surety bond. BATF currently issues a two-year renewable permit.

Age Restrictions. The BATF has no formal age restrictions on who can and cannot manufacture and use fuel alcohol. However, some states do require that you be twenty-one years of age or older, while others are content with a minimum of eighteen years.

License Fee. The BATF has no fee for licensure. Federal law allows you to make 10,000 proof gallons per year without having to post a bond. If you make over the maximum, the bond required is based on a per-proof-gallon basis. If a bond is necessary, the BATF will supply you with a list of approved surety bonding companies, and with blank bond-application forms. Bond is currently computed at the rate of $21 per proof gallon produced in a fifteen-day period, and rounded off to the next $100.

State-by-state fees vary in those states that require a license; and there is an occasional difference in license-type cost. In Pennsylvania, for example, the fee for an experimental license is only $25, but the fee for a commercial license is $2,500. In Tennessee the fee is based on production. In Montana there is no fee, but you must post a $1,000 bond to go into production. In West Virginia a license is required but no fee. The chart on page 28 will give you an idea of state fees and requirements, but check with your state before going into production, since license requirements and fees may change from time to time.

SPECIAL REQUIREMENTS

Residence. If you rent or lease the property on which you plan to establish an alcohol facility, you must submit written permission from your landlord along with your application for a permit. It can simply state that he/she is aware of your activities and has no objections. For your own protection, place a time factor within the body of the permission (for so long as you shall occupy the property, or for a five- or ten-year period). Once your facility is established, your landlord may have a change of mind and tell you to cease and desist, and remove your equipment. In that instance, you could probably sue for damages, but why go through the bother.

Denaturing. Federal law requires that your alcohol be denatured (made not natural nor fit for human consumption) as it is produced. The old federal law required Completely Denatured Alcohol (CDA), such formula being outside the realm of the average person to reasonably handle. This was primarily designed for industrial use and encompassed not only fuel alcohol but other uses of the CDA as well: manufacture of chemicals, proprietary antifreeze preparations, solvents, thinners, etc. The new law makes it much easier for the average person to denature alcohol by making gasoline (or kerosene) the denaturant. This, however, is restricted to fuel use only.

Section 212.38 of BATF 27CFR, Parts 211–212, Formula 28A, reads: "(a) Formula. To every 100 gallons of alcohol add one gallon of gasoline or automotive gasoline. Kerosene or deodorized kerosene may be used in place of gasoline in the same amounts." This formula is based on 190 proof alcohol at 60°F. By adding one gallon of gasoline (or kerosene) to 100 gallons of a lower-proof alcohol, you will produce a higher proportion of gasoline to alcohol, and thus exceed federal requirements. (See Denaturing, Chapter Nine: Distillation.)

Monthly Reports. States with license laws generally require that you maintain records of production–alcohol and effluent volume–and some require monthly reports. Regardless of your state's requirements, it is best to maintain daily (or weekly) records of your activities in a bound ledger. You will at least need bimonthly records to satisfy federal bond requirements; to show that you are not producing more than 10,000 proof gallons of fuel per year or to show that you are complying with your original estimate of production. And, if you are ever investigated, accurate and neat records will be to your advantage.

Taxes. To encourage fuel alcohol production and use, the federal government exempts fuel alcohol (gasohol) from the federal excise tax of 4 cents imposed on gasoline. Unfortunately, this savings is sometimes offset by a state excise tax. In West Virginia, for example, a per-gallon excise tax of 10½ cents is imposed on anything (methanol or ethanol) mixed with gasoline to produce gasohol for sale or distribution. The state tax on fuel alcohol sold for motor fuel use is a median 8 to 9 cents per gallon throughout the country. Check with your state department of revenue regarding your individual situation. Production for personal use is generally exempt from taxation, both federal and state.

Security. Federal law requires security for your equipment (still) and adequate storage facilities for the finished product. This does not mean barbed wire, floodlights, attack dogs, sirens, mine fields, and machine-gun nests, but rather that reasonable precautions be taken against misuse of your equipment and theft of your product. A shed or fenced-in area secured with lock and key is sufficient. Storage in your basement or garage (if lockable) is adequate. The fact that you are denaturing the alcohol as it is produced is itself a deterrent to theft for consumption, but use of your equipment during your absence must also be deterred.

For storage purposes, fuel alcohol in South Dakota falls within the same category as gasoline (Class I Flammable Liquid) and the same rules apply (National Fire Protection Association Standard #30). Alcohol is not nearly so volatile as gasoline, but your state may have stringent storage requirements. Check it out with your state fire marshall or alcohol regulatory agency.

Transporting. You can make fuel alcohol, and you can store it, and even transport it (after it is denatured), but you cannot sell it or transport it for transfer to another individual without special permits and compliance with special regulations. The sale or transfer of alcohol to another individual places you in commercial status–a position you may not want to be in.

COMMERCIAL PERMITS

Why bother knowing the basics of commercial permits if you plan only small-scale production for your personal use?

1. Someday you might decide to go large-scale and start your own business.
2. You may make more than you really need and want to sell your surplus.
3. You may decide to form a co-op fuel alcohol organization of friends and neighbors, or decide to take on a partner.
4. Awareness of commercial permit procedure and regulations makes it easier to live with and comply with an experimental permit.
5. Certain regulations designed for commercial operations may eventually–if not currently–apply to your personal-use operation.

If you are seriously considering going commercial, contact your regional BATF office and request the package of forms necessary to apply for a Commercial Distilled Spirits Permit and the laws governing the permit (Distilled Spirits Tax Revision Act of 1979; Public Law 96–39). After reviewing the forms, material, and regulations, you may change your mind about going big time. Here are the basics:

Personal Questionnaire. This really gets personal. Were you ever known by another name, ever arrested for violation of any federal or state law (traffic violations do not count), ever refused a license related to alcohol manufacture, use, sale, etc.? You must provide: employment history for the past ten years; names and addresses of five references, one of which must be a bank; your residences for the past ten years; and the dollar amount of your interest in the operation and the source of the funds.

How to Apply for an Operating Permit. This defines the operations to be performed–distilling for industrial use, denaturing spirits, warehousing spirits for industrial use, bottling and/or packaging spirits for industrial use, warehousing for nonindustrial use, and any other operations–and manufacturing articles (what will you be using to make the alcohol).

Bond. The guideline for a commercial operation is the same as for the experimental permit: less than 10,000 proof gallons per year requires no bond. However, you can almost be assured of a BATF field inspection of your commercial operation regardless of your production volume.

Environment. Part of the required information relates directly to the environment: the type heat and power you will be using, the type anti-air-pollution equipment you will be using; the amount and composition of all solid waste material anticipated (mash and by-products) and liquid wastes generated; and planned disposal. If you plan to use an incinerator for disposal, what type anti-air-pollution equipment will you be using? What type operational noise sources are anticipated?

The Environmental Protection Agency (EPA) is promulgating air and water pollution standards as they apply to the manufacture of fuel alcohol. These standards, however, are basically designed for commercial operations and should have little impact on the small-scale operator. Dependent on the size of your operation, you may be required to obtain a National Pollutant Discharge Elimination System (NPDES) Permit. State and local pollution control regulations will probably have more effect on your operation than federal standards. Be on the safe side; check it out with EPA.

If you plan to discharge waste into a navigable body of water, a pond, or an underground water table, you must obtain state certification, and federal certification under the Federal Water Pollution Control Act, Section 21(b) of Public Law 91-224, Form 4805 (1740.2).

Generally, as far as the BATF is concerned, they will not investigate nor involve or notify the Environmental Protection Agency unless your plans represent an obvious threat to the environment.

Occupational Safety and Health Act (OSHA). Whether small-scale private operation or large-scale commercial, if you employ more than ten people, you come under OSHA regulations. This can be one big headache, for compliance with OSHA is very complex. To obtain your regional or local OSHA toll-free number, dial: 1-800-555-1212. A quick telephone call will give you compliance information.

Other Considerations. Your commercial operation most probably will have to comply with several other standards as well. Alcohol manufacture and storage is classified as Class I, Group D in the National Electrical Code. Grain handling and grinding is considered Class II, National Electrical Code, Article 502. This means that your electrical system may need upgrading to comply—wiring that is heat and chemical resistant to within tolerances set by the code, and special receptacles, fixtures, and junction boxes that diminish the possibility of a spark causing an explosion. To your operation, this means that if you build your own, it had better be good. To be on the safe side, contact your state fuel alcohol regulatory agency. (See Contact Information, State-by-State Guide to Fuel Alcohol Information.)

You should also be on the alert for other special state considerations. For example, Louisiana requires that a green dye be added to gasohol to distinguish it from straight gasoline. Tennessee requires addition of a yellow dye. While this is no major process and is not expensive, you should be aware of your state's regulations for legal compliance.

FEDERAL SEARCHES AND INSPECTIONS: WHAT TO EXPECT

Once you have obtained a fuel alcohol permit—whether experimental or commercial—BATF has the right to enter your premises without warrant, dig up ground and rip into walls if concealment of pipelines or material is suspected, and basically invade your privacy at their discretion. Those states with license laws generally have the same rights, with at least the right to enter your premises to inspect your equipment and other aspects of your operation. It is therefore best to keep your operation completely apart from your actual place of residence with no connection whatsoever of plumbing, sewer lines, or anything else that looks as if it could convey alcohol.

For the most part, the BATF (or state liquor control agent) will neither take the time nor expend the manpower to check out a small operation. It is quite possible, however, that as the number of private small operations increases, states will spot-check operations for compliance with state laws.

Addresses of BATF Regional Offices

CENTRAL REGION
Indiana, Kentucky, Michigan, Ohio, West Virginia

Regional Regulatory Administrator
Bureau of Alcohol, Tobacco and Firearms
550 Main Street
Cincinnati, OH 45202

MID-ATLANTIC REGION
Delaware, District of Columbia, Maryland, New Jersey, Pennsylvania, Virginia

Regional Regulatory Administrator
Bureau of Alcohol, Tobacco and Firearms
2 Penn Center Plaza, Room 360
Philadelphia, PA 19102

MIDWEST REGION
Illinois, Iowa, Kansas, Minnesota, Missouri, Nebraska, North Dakota, South Dakota, Wisconsin

Regional Regulatory Adminstrator
Bureau of Alcohol, Tobacco and Firearms
230 South Dearborn Street, 15th Floor
Chicago, IL 60604

NORTH-ATLANTIC REGION
Connecticut, Maine, Massachusetts, New Hampshire, New York, Rhode Island, Vermont, Puerto Rico, Virgin Islands

Regional Regulatory Administrator
Bureau of Alcohol, Tobacco and Firearms
6 World Trade Center, 6th Floor
(Mail: P.O. Box 15, Church Street Station)
New York, NY 10008

SOUTHEAST REGION
Alabama, Florida, Georgia, Mississippi, North Carolina, South Carolina, Tennessee

Regional Regulatory Administrator
Bureau of Alcohol, Tobacco and Firearms
3835 Northeast Expressway
(Mail: P.O. Box 2994)
Atlanta, GA 30301

SOUTHWEST REGION
Arkansas, Colorado, Louisiana, New Mexico, Oklahoma, Texas, Wyoming

Regional Regulatory Administrator
Bureau of Alcohol, Tobacco and Firearms
Main Tower, Room 345
1200 Main Street
Dallas, TX 75202

Addresses of BATF Regional Offices

WESTERN REGION
Alaska, Arizona, California, Hawaii, Idaho, Montana, Nevada, Oregon, Utah, Washington

Regional Regulatory Administrator
Bureau of Alcohol, Tobacco and Firearms
525 Market Street, 34th Floor
San Francisco, CA 94105

State License Law Chart

The following states require a license or registration for fuel alcohol manufacture for personal use. If your state is not listed, it does not automatically mean that your state does not require a license. Check with your state's fuel alcohol control board before you begin production.

State	*License*	*Fee*
ALABAMA	YES	$100.
ARKANSAS	REGISTRATION	NONE
CALIFORNIA	YES	$13.20
DELAWARE	YES	NOT SET
IOWA	YES	$350
MONTANA	YES	$1,000 BOND
NEW YORK	YES	$625
NORTH CAROLINA	YES	$10
OKLAHOMA	YES	$25
PENNSYLVANIA	YES	$25
SOUTH DAKOTA	YES	NOT SET
TENNESSEE	YES	BASED ON PRODUCTION
TEXAS	YES	$100
UTAH	SPECIAL PERMIT	NONE
WEST VIRGINA	YES	NONE
WISCONSIN	YES	$10

3
Feedstocks

SINCE ALCOHOL is produced by the fermentation of simple sugars, principally glucose, it is imperative that you have access to an adequate supply of feedstock that either contains sugar or contains carbohydrates that can be converted to sugar. With the exception of forestland (cellulose, see Chapter Six: Microorganisms), you must then turn to agricultural products (crops) as your feedstock source. The two most important questions you must ask yourself are: Do you grow your own or buy your feedstock, and which feedstock is best for you and your operation?

GROWING YOUR OWN

Being able to grow your own depends on where you live and land availability. If you have a small back yard just big enough to erect a still, growing your own is not for you. However, if you have several acres that can be cultivated, it is then worthy of consideration. If nothing else, you can supplement what you buy by growing an acre or two of your own.

Growing your own takes a little time and effort, and a modest investment, so that time must be converted to money—how much is your time worth—added to the cost of seeds and equipment, and compared to the cost of buying. A criterion here, too, is crop yield—how much will you get out of what you put in—in terms of produce and alcohol per acre.

The cost of fuel has more than tripled in the past decade. Fortunately, the cost of a bushel of grain has not. So, as long as the feedstock cost remains relatively low, it is feasible to buy your feedstock and make your own fuel—the cost of the feedstock being offset to some degree by inflated fuel costs. However, as the alcohol movement becomes more widespread and firmly entrenched as a way of life, supply and demand may dictate ever-increasing crop prices, whether you like it or not. It is best to at least plan for that possibility.

Commercial alcohol manufacturers are literally at the mercy of the market. With fifty to one hundred million dollars invested in a sugar-beet-based alcohol plant, the commercial operator can hardly switch to using grain if the price of sugar beets skyrockets. In this one respect, you have a definite advantage over the large-scale operation. You can switch from one feedstock to another with relative ease and low cost.

If you grow your own, what should you grow? If you buy, what should you buy? Naturally you want the highest alcohol yield per acre planted, but some crops are definitely not going to be within your realm. Sugar cane, for example, is grown successfully in only four states: Florida, Louisiana, Texas, and Hawaii, as well as in the commonwealth of Puerto Rico. If you do not live in one of those states, sugar cane is not for you. Other crops, such as corn, sugar beets, and potatoes, can be grown just about anywhere you can grow outdoor flowers. Even so, which is best for your operation?

The following table of selected crops will help you begin your decision-making process. It is based on crops that have a high carbohydrate content (about 60 percent), and how much harvest you can expect per acre planted (yield per acre).

CROP	*% CARBO-HYDRATE*	*ACRE YIELD AVERAGE*	*APPROXIMATE ALCOHOL YIELD/ACRE*
Beans	61–64%	1500 lbs/acre	71 gals/acre
Wheat	70–75%	1804 lbs/acre	99 gals/acre
Field corn	72.2%	6104 lbs/acre	334 gals/acre
Split peas	62–63%	1500 lbs/acre	71 gals/acre
Rice	77–80%	4500 lbs/acre	268 gals/acre
Rye	72–73%	1456 lbs/acre	80 gals/acre
Sorghum	70–72%	3472 lbs/acre	187 gals/acre

The percent carbohydrate factor is based on an average of all types of beans, wheat, etc. The acre yield average will vary tremendously throughout the United States. For example, the average field corn yield per acre runs from a low in Florida of 53 bushels, to a high in Illinois of 128 bushels. For more accurate crop yield statistics for your state, contact your state department of agriculture.

The approximate alcohol yield per acre is based on the axiom that yeast will, under the most ideal conditions, ferment half of the total weight of fermentable matter in a given feedstock on one fermentation run. This figure is always open to debate, is in constant controversy, and is often grossly misquoted. However, for the sake of comparison, as long as the same factors are used throughout, the end result will show which is basically the most favorable feedstock for your operation. The basic formula for any feedstock considered is to multiply the percent carbohydrate times the pounds per acre yield of the crop, which gives you the total weight of fermentable matter. Multiply that by 50 percent to obtain the weight of potential alcohol from that feedstock. Multiply that by 6.6 pounds per gallon to obtain the number of potential gallons of alcohol. Field corn example: 72.2% × 6104 = 4407 lbs × .50 = 2203.5 ÷ 6.6 = 334 gals/acre.

Sugar beets and sugar cane do not fall within this formula, for they are not basic starch crops. The fermentable matter is the natural sugar within the beet or cane. The percentages of sugar in cane or beets will vary depending on region, growing conditions, and when it last rained. Rainfall, or the lack of it, will affect the sugar content by as much as two percent, which is more than 10 percent of the total sugar content of the feedstock.

CROP	*% SUGAR*	*ACRE YIELD AVERAGE*	*APPROXIMATE ALCOHOL YIELD/ACRE*
Sugar Beets	14–18%	19.6 tons/acre	398 gals/acre
Sugar Cane	12–14%	38.4 tons/acre	652.8 gals/acre

Based on the 10 percent variable mentioned above, the gallons-per-acre yield could drop to 360 and 587, respectively.

According to the National Gasohol Commission, the alcohol yield per ton of sugar beets is 20.3 gallons; for sugar cane it is 17 gallons. To determine your potential alcohol yield per acre, you would multiply the gallons per ton given above by your actual crop yield in tons per acre. This does not take into account the ten-percent sugar-content variable.

Potatoes are a major crop that can be easily grown almost anywhere. The average yield per acre is 272 cwt (hundredweight) or 27,200 pounds. The average alcohol yield is 28.8 gallons per ton, or about 392 gallons per acre, according to the National Gasohol Commission.

Actual per-acre alcohol yields will always depend on actual crop yields. The following table offers a profile of average major crop yields.

Average United States Crops Yields Per Acre (Source: USDA)

CROP	*AVERAGE U.S. YIELD*	*HIGH*	*LOW*
Barley	50.6 bu	75 bu	39 bu
Beans	1457 lbs	1930 lbs	400 lbs
Corn	109.4 bu	128 bu	53 bu
Grain sorghum	62.9 bu	75 bu	37 bu
Hops	1727 lbs	1800 lbs	1290 lbs
Oats	54.4 bu	70 bu	39 bu
Potatoes	27,200 lbs	47,500 lbs	9,000 lbs
Rice	4,588 lbs	6,450 lbs	4,050 lbs
Rye	25.9 bu	32.2 bu	17 bu
Soybeans	32.2 bu	38.5 bu	23 bu
Sugar beets	19.6 tons	26.8 tons	15 tons
Sugar cane	38.4 tons	98.1 tons	20.5 tons
Sweet potatoes	11,500 lbs	16,500 lbs	7,500 lbs
Wheat	34.2 bu	76.3 bu	21.4 bu

To convert pounds to bushels, divide by 56. To convert bushels to pounds, multiply by 56. If you grow your own, you can expect a crop yield somewhere between the lows and highs indicated. For more accurate local information, contact your state department of agriculture, since crop yields are subject to a variety of conditions: geographic location, climate, and soil conditions to name a few.

Another major concern is raw material cost recovery, or Distiller's Dried Grains (DDG). The following Department of Energy cost recovery table offers a profile of the most common DDG/raw material cost factors. Bear in mind that these figures are based on a commercial operation, which requires drying the material after use and prior to delivery to the farmer. On a small scale, you can save the cost of drying by delivering wet residue to the farmer (his livestock will eat it wet or dry). In general, wet residue can be stored up to four days before using. Dry residue can be stored almost indefinitely.

Whether you plan to use or sell your fermentation residue wet or dry, there are a few things to consider.

Fermentation creates various acids of varying strengths and potencies, which, regardless of the type distillation utilized, remain in the effluent and residue by-products. Livestock consuming the residue are acid tolerant to a point, which means that they can and will eat only a certain amount of acidic residue. Like humans, animals are not too fond of heartburn and indigestion. Animals cannot go to a drugstore and buy an anti-acid pill or liquid, so they solve their problem by avoiding the food source. It may, therefore, become necessary for you to neutralize the fermentation residue to a pH of 7 to encourage your livestock to eat more of it. This is an individual situation, and really depends on your particular livestock (cows, pigs, or chickens) and fermentation feedstock (corn, sugar beets, potatoes, etc.).

Mold and bacteria create residue problems, more so with the wet than the dry. Although not a total solution, this problem can be somewhat diminished by exposing the residue to heat and/or sunlight. It is only natural to assume that you can kill mold or bacteria by using a food preservative or antibiotic. However, if it is not now illegal in your state, it soon will be. Preservatives, such as the propionates or benzoates, and antibiotics, such as penicillin, tetracycline, and streptomycine, used on livestock feed will pass on to the animal, and eventually pass on to the consumer of the slaughtered livestock. This could have an adverse effect on the consumer, particularly if the individual is allergic to the antibiotic used. Another detrimental aspect of using preservatives or antibiotics is that the animal may develop adverse reactions to certain chemical agents; the animal's immunological system may be adversely affected; or certain bacteria common to certain animals will become immune to antibiotics so that when an animal becomes ill, it will not be possible to treat it back to good health.

bu = bushel *T = ton*	*Corn*	*Grain Sorghum*	*Wheat*	*Potatoes*	*Sugar Beets*	*Starch*	*Molasses*
Cost raw material ($):	2.50/bu	2.30/bu	3.35/bu	20.00/T	26.00/T	.08/lb	.36/gal
Yield in gallons:	2.60/bu	2.60/bu	2.60/bu	28.80/T	20.30/T	.06/lb	.40/gal
Per gallon cost ($):	.96	.88	1.29	.69	1.28	1.20	1.00

bu = bushel *T = ton*	*Corn*	*Grain Sorghum*	*Wheat*	*Potatoes*	*Sugar Beets*	*Starch*	*Molas*
By-product sale ($):	110/T	110/T	110/T	6/T	93/T	0	15/
By-product credit/ gallon ($):	.36	.36	.44	.03	.60	0	.12
Net material cost/ gallon ($):	.60	.52	.85	.66	.68	1.20	1.1

Before you make a feedstock decision you should chart it as above, based on current and local market conditions. Even if you grow your own, the raw material cost recovery should be computed into your bottom-line decision.

WHICH CROP TO USE

Starch Crops. Of the starch crops, corn is the most popular, abundant, and—second only to grain sorghum—cost-feasible feedstock. However, market conditions can change the cost status almost on a month-to-month basis. Corn is easily homegrown, and is available for purchase in every state. Corn offers the greatest potential for bartering with a farmer—let me borrow your corn for a week and I'll return it 17 percent richer in protein—since almost every farmer with livestock stores corn in silos for feed. The price will generally remain stable, for there are over seven billion bushels of grain corn produced in the United States each year, compared to less than one billion of grain sorghum.

In what form should you buy the corn—whole kernel, cracked, or ground (cornmeal)? You will need one eighth of your feedstock in whole kernel form for sprouting or malting. (See Chapter Seven: Mashing). The other seven eighths can be purchased cracked or ground; or you can buy whole kernel and have it cracked or ground at a mill; or buy whole kernel, cook it, and grind it yourself. The major advantage of buying all whole kernel is that it is less expensive than cracked or ground. Also, most farmers with whom you might deal will be selling whole kernel right from their storage facilities.

Other Grains. Wheat, barley, and rye are high-cost, low-yield grains, and are often used by moonshiners to flavor their end product. Various quantities (formulas) added to corn mash will vary the taste and mellowness of corn "likker." For the novitiate fuel alcohol producer, these grains are not that great unless: the supply is tremendous and the cost exceptionally low; or if occasion offers a few tons of spoiled grain that can be cheaply bought, keeping the cost per gallon of alcohol made below that of corn-made.

Sugar Crops. Sugar beets are most favored because of the ease of growing the beet and processing it into alcohol. The sugar-extraction process is not that complicated, so the average back-yard enthusiast can make sugar beet alcohol start to finish with little difficulty. Sugar beets have two major disadvantages. The first is the cost versus yield per gallon of alcohol, if you buy the beets. The second is that beets are limited to a single mash/distillation run, whereas grain can be run several times until almost all of the starch has been depleted. If you grow your own (and beets can be grown just about anywhere), the higher market cost is somewhat offset because of the high yield per acre in both beet product and gallons of alcohol. The sugar beet root grows to about two feet in length, breaks off and remains in the soil when the beet is harvested, and provides an excellent source of nitrogen for the soil. It also provides humus, which aerates the soil and retains water. It is ideal as a rotation crop—grow corn one year and beets the next. Beets can be used either fresh or dried, although drying depletes some of the sugar content. Dried beets can be stored almost indefinitely, so you can stock up when the price is right, or grow more than you really need and store the rest.

The fodder beet is currently being cultivated in New Zealand. It is a cross between two species of beet—the sugar beet and mangold. It offers a higher sugar yield per acre, is hardier and more easily grown, and loses less sugar during storage than normal sugar beets.

The Jerusalem artichoke is not really a true artichoke, but it is an emerging and promising sugar crop. Most of the sugar is found in the stalk, although some finds its way into the tuber and root system. The Jerusalem artichoke is a perennial, so tubers left in the ground will sprout the following year, which greatly offsets the time and expense of planting each year.

Sugar cane is ideal for commercial operations, but not too suited for the do-it-yourselfer. Sugar cane must be crushed and cooked and the sugar must be extracted and subsequently fermented. The raw-material cost is higher than that for corn or sugar beets, and the cost and complexity of the processing equipment involved generally prohibits sugar cane as a feedstock. On the plus side, the per-acre alcohol yield is extremely high—fully one third higher than that of sugar beets—and the by-product, called bagasse, can be used in a variety of ways: burned for fuel, used as fertilizer, or used in the manufacture of fiberboard and front-door welcome mats.

A problem consideration with all sugar crops is storage. Moist sugar is readily attacked by microorganisms–yeast, bacteria, and mold. If the harvested product is not used right away, steps must be taken to obviate the effects of microorganisms. This can be done in two ways: dehydration and concentration. Dehydration is simply drying the product. Without moisture, microorganisms will not develop. This is not feasible for sugar cane. Concentration is extracting the sugar from the product, and then concentrating it by cooking and removing water to a sugar percentage in excess of 50 percent. The higher the sugar content above 50 percent, the less chance for microorganism survival in the medium. (You may have noticed that honey can be stored for extremely long periods of time without refrigeration and without preservatives.)

Potatoes. Although potatoes are abundant, inexpensive, and offer a high per-acre alcohol yield, there are a few problems associated with potatoes as a feedstock. Potato starch is different from grain starch, and is more difficult to hydrolyze (convert the starch to sugar). There is very little use and little market for the by-product, so cost recovery is miniscule. In fact, the cost recovery of grains is twenty times that of potatoes. Storage presents a minor problem, for potatoes are not easily dehydrated. Placed on a shelf to dry, potatoes will either rot or grow long shoots. Potatoes are not infeasible, simply less feasible than other feedstocks.

Fruits. The highest and best use for fruits is for human consumption, either by eating or drinking (wine). Unless you have access to distressed fruit or there is a regular surplus in your area, do not rely upon fruits as your primary feedstock. The cost, compared to other feedstocks, is just too prohibitive for consideration. Those not suited for human consumption must be used to make ethanol very shortly after beginning to turn bad or rot. Lack of year-round availability is a definite hindrance. Nevertheless, check fruit growers around harvest time for distressed produce.

Restaurant Slop. Feasible? Yes. A lot of trouble? Yes. If a restaurant is willing to segregate waste vegetables from meat and other table scraps, slop is a viable feedstock. It is especially feasible if you know a hog farmer willing to take the slop after you have fermented it. If you must pay the restaurant owner for the slop, then charge the farmer the like amount. If the farmer went to the restaurant, he would have to pay something for the slop anyway. The real problems with slop are bacteria, smell, and handling. This is somewhat offset in the winter when cold weather inhibits mold and bacteria growth. An advantage to using slop is that part of the cooking for hydrolysis has already been done for you. The slop will also contain a certain amount of sugar that does not need to be hydrolyzed. However, there will also be small amounts of salt and vinegar, either of which will inhibit fermentation. If the salt content exceeds 15 percent, the mash must be diluted with water to decrease the total percentage of salt. If vinegar is obviously present (you will smell it) calcium carbonate can be added to the mash to buffer the effects of the vinegar.

Supermarket. Check supermarkets periodically for any waste vegetables or fruit they may be disposing. Every food store has its share of spoiled food, which it cannot sell to the general public. Most times they simply throw it away and suffer their losses. If you are willing to haul it away, they should be willing to cooperate.

Halloween. This is an annual occasion that provides a literally free feedstock in two ways. Firstly, pumpkins are generally discarded within a week after Halloween, and the whole pumplin–shell, pulp, and seeds–will ferment. Pumpkin seeds are ideal for fermentation when dried and pulverized. Secondly, children usually collect more candy than they can possibly eat. Parents end up discarding much of their children's Trick-or-Treat acquisitions several weeks after Halloween. With the exception of chocolates, most candies are extremely high in sugar content. Candies can be dissolved in water and fermented, or simply added directly to a mash run–corn for example–to raise the sugar level and enhance fermentation. Your cost is basically the time it takes to let people know you are interested in their discards, and to make the eventual collection.

In conclusion, you can ferment anything that contains either sugar or carbohydrates that can be broken down to simpler sugars. You can mix or match fermentable feedstocks–a little corn, a few potatoes, some fruit or other vegetables–they all work together to make fuel alcohol.

4
Layman Chemistry

FUEL ALCOHOL production involves processes including both organic (that which lives and contains carbon) chemistry, and inorganic (that which is not alive, and contains no carbon) chemistry, as well as physics (distillation and the mechanical application of the end product). Chemistry and physics are generally shrouded in a cloak of formulas and mysterious terminology that baffle and confuse the average person. By the time most people finish reading a paragraph loaded with Latin and dotted frequently with symbol equations, they are no closer to understanding than when they started. It need not be that way. Chemistry and physics can remain as simple as learning the terminology involved. The basic principles involved–actions, reactions, freezing and boiling points–are just that: basic and easy to grasp.

True, you can make alcohol by using a basic mash formula and following a step-by-step process, and distill alcohol without understanding what is happening and why. But if you don't understand why something is happening, you can hardly avoid mistakes; and when something does go wrong, you can hardly begin to trace back to the source of the problem. Without a basic understanding you can waste a lot of time trying new methods by trial and error in an attempt to improve your operation or your alcohol yields. And there will be questions as you go along–why is this happening? Natural curiosity.

THE BASIC BASICS

Different chemicals react in different ways, obviously. But some chemicals can be both beneficial in some respects and detrimental in others. If you add salt to your food, it enhances the flavor. Add one teaspoon of salt to one-half glass of clear soda pop, and it *will* fizz and it *will* soothe your upset stomach or heartburn. Salt is used as a preservative, and in ancient times it was applied to open wounds to inhibit infection; very painful, but at least the patient had a better chance of survival. Conversely, salt is deadly to yeast. In concentrations exceeding 10 to 15 percent, it will rupture the cell wall and kill the yeast. This is detrimental to the yeast, and so, detrimental to your alcohol-making endeavor. But what is salt, really?

Salt is sodium chloride with the chemical formula NaCl; Na = sodium, and Cl = chlorine. Taken alone, sodium has none of the characteristics of salt other than being a solid. Chlorine is entirely dissimilar to salt, for it is a poisonous gas. When chemically combined, they form NaCl, and give up their original personalities.

Chemical and Mechanical Reactions. Add salt (NaCl) to water (H_20) and the salt dissolves, seeming to disappear. Boil the solution until the water has all vaporized, and the salt is left behind as a residue. This is a basic mechanical reaction; no chemical changes have occurred—the salt is still salt, and the water is still water, even when vaporized to steam. The bond between sodium and chlorine is just as strong as the bond between hydrogen and oxygen (H_20), and none wants to change sides—or chemically react. The same reaction would occur if you added sugar to water. It would dissolve, and when the water was removed, it would precipitate back to sugar. Add salt and sugar to water, remove the water, and you are left with a mixture of salt and sugar, but still no chemical reaction has taken place.

If you add sugar to sulfuric acid (H_2SO_4 = hydrogen, oxygen, and sulfur), you will see a distinct chemical reaction. The sugar will fizz, bubble, and boil, emitting gas as it does, and the sugar will be converted to a lump of black carbon. Once this reaction has taken place, there is no way you can convert the lump of carbon back to sugar again—neither mechanically nor chemically.

With a mechanical reaction, there is usually a change of form, but no change in the chemicals involved—they are simply temporarily altered. Generally, if chemicals are mechanically combined, they can be mechanically separated back to their original state. With a chemical reaction, there is usually a change of form, and a change of chemical characteristics—new chemicals are formed. When chemicals are chemically combined, the reaction is either reversible or nonreversible. Reversible means that once a chemical has been altered or changed to one form or new composition, the addition of another chemical or chemicals can change it back to its original state. A nonreversible reaction, as with the sulfuric acid and sugar above, is a one-way street; once the reaction occurs, that's it. So what causes chemical reactions?

ATOMS AND MOLECULES

Although the neutrino is now accepted as the smallest particle of matter known to man, for the sake of convenience, the atom shall be considered the smallest particle of any significance in this book—both in definition and usage.

With the exception of the hydrogen atom, all known atoms consist of a nucleus containing neutron(s) and proton(s), surrounded by a ring(s) of electron(s). An atom can best be imagined by picturing our solar system, with the sun as the nucleus surrounded by the planets as the electrons. Just as the planets are not all equidistant from the sun, so electrons around certain atoms form layers, or shells. The electrons in the outermost shell are the ones that do the actual combining in the formation of compounds and functional chemical reactions. In fact, it is the outermost shell of electrons that actually holds the atom together.

A *molecule* is comprised of one or more similar atoms; has a neutral electrical charge—neither positive nor negative; and is the smallest particle of an element that still retains the characteristics of that element. If there is a charge—positive or negative—to the atom(s) it is not a molecule, it is an *ion*. A positive ion has a deficiency of electrons; a negative ion has an excess of electrons.

A *compound* is comprised of two or more different atoms. Compounds containing negative ions also contain positive ions; for example, table salt (NaCl) has a positive sodium ion (Na^+) and a negative chlorine ion (Cl^-). This attraction between positive and negative is known as electrovalence. This is the primary bonding method of inorganic chemicals—an attraction between positive and negative charges. With organic chemistry, the primary bonding method is covalent, which is a filling out of the electrons in the outer shell of the atom. Carbon, which is the real foundation of organic chemistry, has four electrons in its outer shell. However, it wants eight and will do almost anything to get the other four. This is to say that carbon will combine with just about anything to make its shell complete. Almost all organic chemical reactions are covalent in nature.

Hydrogen is the favorite bonding element for carbon. Hydrogen has only one electron, so four hydrogen atoms are required to complete the outer shell of the carbon's eight. This simple bond is called a hydrocarbon, and it looks like this:

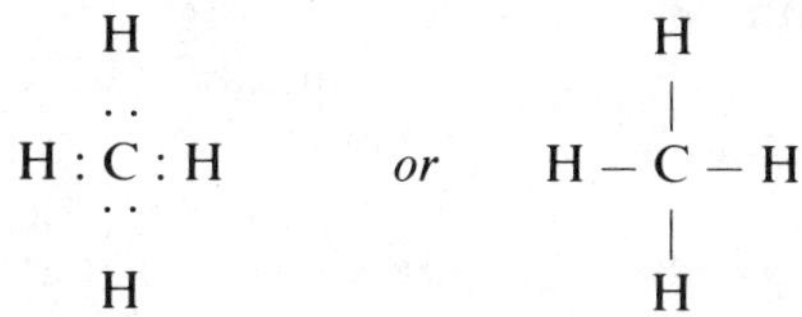

$= CH_4$ = Methane

Each dot represents an electron—four surrounding the carbon atom, and one surrounding each hydrogen atom—paired up around the carbon to complete the outer shell's eight.

Oxygen is the next favorite element for filling outer shells, and bonds about as easily with hydrogen as with carbon. The true composition of a chemical, however, is determined by where the bonds occur, and by the bonding characteristics of a given reaction or compound.

Bonding Sites. Carbon always requires four bonds to complete itself; oxygen always requires two bonds; and hydrogen always requires one bond. But it isn't as simple as that. There are single bonds, double bonds, and triple bonds, depending upon the particular chemical or reaction, arranged in either an open chain form or closed loop form. Of significance to alcohol production are sugar and starch chains.

Glucose (sugar), for example, is a six-carbon, straight-line chemical, with six oxygens and twelve hydrogens. The formula is: $C_6H_{12}O_6$, and looks like this:

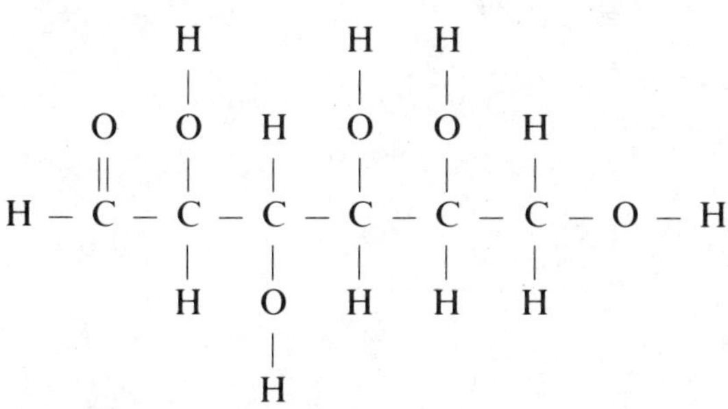

= Glucose = Monosaccharide

You will notice, if you count the bonds, that each carbon atom has four bonding sites; each oxygen, two bonds; and each hydrogen, one.

Monosaccharides are the simplest forms of sugars, meaning that they cannot be broken down into simpler sugars. To form a disaccharide, you simply combine two monosaccharides at the hydrogen/oxygen end to form a longer chain. Sucrose is the most commonly recognized disaccharide, and is formed with one molecule each of glucose and fructose, another monosaccharide. Going one step further, and forming longer chains, are polysaccharides, which are the starches and cellulose.

The real significance of bonding, chains, and saccharides is that you will find them all ready-made in nature. However, creating the bonds and chains is not your concern. To make alcohol, you must break the disaccharides and polysaccharides down to simple, fermentable monosaccharides. This will be covered more thoroughly in Chapter Seven: Mashing.

FUNCTIONAL GROUPS

Throughout chemistry you will find *functional groups.* These are various combinations of—but not restricted to—carbon, hydrogen, and oxygen, just hanging around waiting for something to happen. Like a cake mix that is all set to go when water is added, functional groups are all set to go under the right circumstances. Functional groups usually step in when a hydrogen atom steps out. When the hydrogen departs, it leaves a bonding site. That reactive section is known as an alkyl group and is signified by the letter *R*. Alcohol is represented by the functional group: R equals OH. Ethanol looks like this in action:

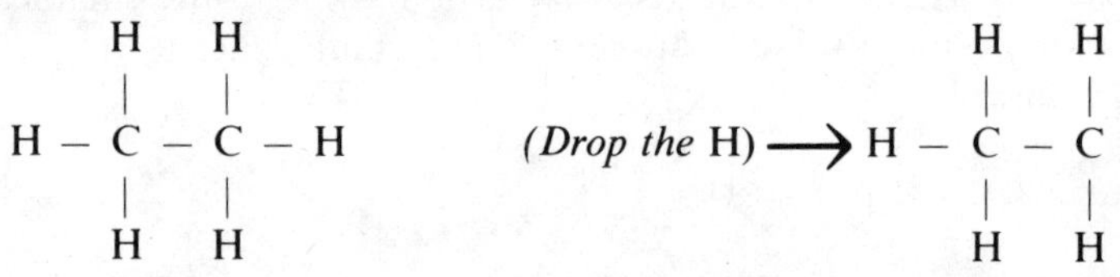

Ethane *Alkyl Group*

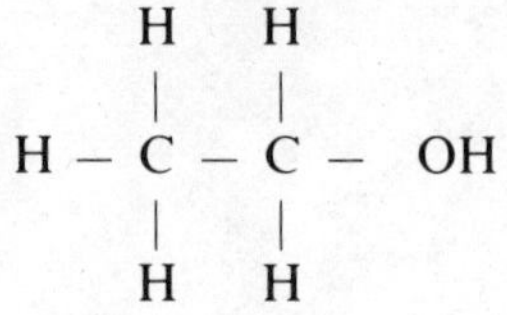

Ethanol *(Add the* OH) ⟶

You may see the ethanol formula written: CH_3CH_2–OH, or Ch_3CH_2–R where R equals OH, or C_2H_5–OH, all of which, when you count the carbons, hydrogens, and oxygen, add up to the same thing: ethanol alcohol (2 carbons, 6 hydrogens, and 1 oxygen).

If the alcohol-producing functional group is not readily available, another group will step in, such as aldehyde, and it looks like this:

Aldehyde Functional Group: R equals 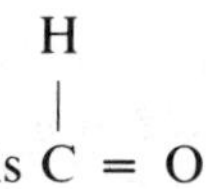

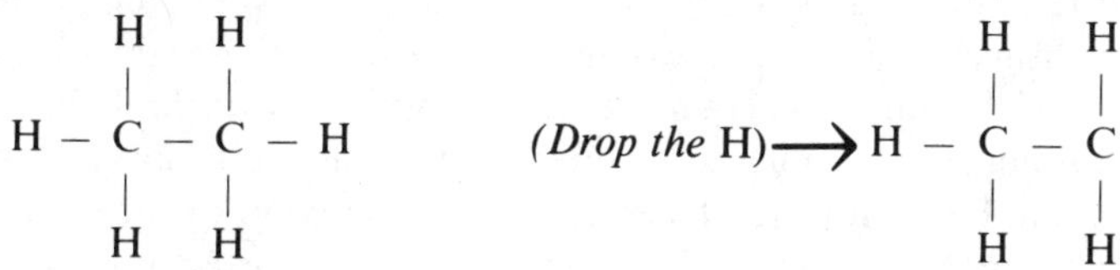

Ethane *Alkyl Group*

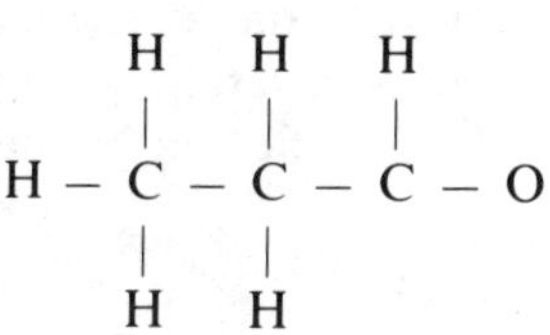

Propaldehyde

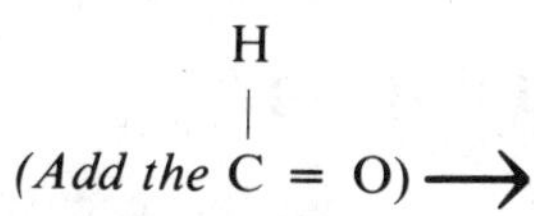

Functional groups will bear more significance in the next chapter on sugars.

CHEMICAL REACTIONS

Oxidation/Reduction. The most common reactions with which you will be concerned are oxidation and reduction. Oxidation is most commonly thought of as the action of oxygen on an element or compound, with a resultant change of form or chemical composition; for example, the rusting of metal. By the truest chemistry definition of the word, oxidation is a loss of electron(s), and reduction is a gain of electron(s). The simplest way to remember the two is with the acronym LEO GER. *L*ose *E*lectron = *O*xidation, *G*ain *E*lectron = *R*eduction. These reactions are generally carried out using hydrogen donors and acceptors; and there is an intrinsic balance to the system—as one compound loses an electron to an acceptor it subsequently gains an electron from a donor. During fermentation, a series of oxidations and reductions occurs, yielding different compounds along the way, until the final product is realized—acetaldehyde—which is either oxidized to acetic acid or reduced to alcohol.

Hydrolysis. True hydrolysis is the action of water ions ($H_2O = H^+$ and OH^-). The major role of hydrolysis in alcohol production is to hydrolyze (divide or split) disaccharide and polysaccharide chains into simple, fermentable sugars. In more simple terms, a water molecule cleaves the disaccharide or polysaccharide into simple sugars. With sucrose, a disaccharide, a water molecule simply cuts it in half. With starch or cellulose, both polysaccharides, it may cleave at any point along the chain, but will generally work in from one end or the other. Eventually, the polysaccharide chain will yield the shorter monosaccharide chains necessary for fermentation.

Saccharide hydrolysis can be accomplished in several ways: prolonged boiling of the saccharide in water, boiling with the assistance of enzymes, and boiling with the assistance of one of several types of acids. Details of hydrolysis will be covered in Chapter Seven: Mashing.

ACIDS AND ALKALIES

Acids and alkalies are highly reactive and corrosive compounds that can alter the course of a chemical reaction either positively or negatively. During fermentation, small amounts of acid salts and compounds are formed naturally. In fact, yeast prefers a slightly acid environment. However, too much acid will adversely affect yeast, killing it or rendering it dormant and useless.

Acids and alkalies can be diluted with ordinary water—the more water, the weaker and less reactive the acid or alkali. The most common acids used in alcohol production are sulfuric acid (H_2SO_4), and hydrochloric acid (HCL), also known as muriatic acid. Either can be used for hydrolysis, although hydrochloric acid is more readily available—most hardware stores sell it. However, H_2SO_4 is most effective. The most common alkalies are lime/quicklime (calcium oxide, CaO), and sodium hydroxide (NH_3OH). *Both acids and alkalies must be stored, handled, and used with extreme caution,* for both have the ability to cause severe skin burns and the potential to cause blindness. Both can cause chemical pneumonia and/or lasting lung damage.

Acids will always be in liquid form, and are sold in either glass or plastic containers. Once purchased, acid should never be transferred to a metal container for storage, for the acid will react with metal, create hydrogen gas, and eat its way through the container. Acids are sold in either diluted or fuming form. Fuming acid is close to its purest form, and visible fumes are apparent when the lid is removed. Whether you buy diluted or fuming, always use the acid in a well-ventilated area, and *never, ever breathe the fumes.*

When using acids (or alkalies) always wear rubber gloves, preferably elbow length, and safety glasses. Most ideally, wear a full-length, rubberized apron. If acid spills or splashes onto regular clothing, it will eat through the cloth and have contact with your skin. This is almost as bad as spilling gasoline on your clothing and igniting it. By the time you remove the clothing, the damage will have been done.

Always have a bucket of water or hose nearby to immediately flush your skin if an accident does occur. Acid burns quickly, and by the time you run to a faucet, it may be too late to remedy the situation.

When mixing acid with water or other solution, always pour the acid into the liquid, never the reverse. Depending on the strength of the acid, heat will be generated during mixing. If you pour water into a container of fuming acid, it could bubble and boil and splash up onto your face. Remember to let the acid eat its way into the liquid, and you will have little problem with mixing.

Alkalies are sold dry, but that does not diminish the adverse effects of skin or eye contact. Skin is generally moist and will activate the alkali soon after contact. Eye damage begins almost immediately on contact because the eye is more moist than skin. Follow the same safety precaution as for acids, and always keep a bucket of water or hose nearby.

Just as you should avoid breathing acid fumes, you should avoid breathing alkali dust. The effects are about as catastrophic, though not as immediately apparent.

Above all else, whether you use one or both, store acid or alkali where there is absolutely no chance of access by children or animals. The consequences are obvious.

pH Factor. The pH is the potency of hydrogen, or the hydrogen-ion concentration, and is used to determine the acidity or alkalinity of a substance or compound. pH is classified on a scale of 1 to 14 (pH 1 to pH 14) with pH 1 being extremely acidic and pH 14 being extremely alkaline; pH 7 (water) is neutral. Since yeast prefers a slightly acid environment, it is important to know the pH of your mash. Too acidic or too alkaline will inhibit fermentation and greatly diminish the efficiency of your operation.

There are three common devices for testing and measuring pH: red/blue litmus paper; pH paper; and a mechanical pH meter.

1. **Red/Blue Litmus Paper.** The least expensive of the three methods, it is also the least efficient. Blue litmus paper will turn red in the presence of acid; red litmus turns blue in the presence of an alkali. But neither will tell you the exact pH of either the acid or the alkali. At best, you can determine neutrality (pH 7), for then neither would change color.
2. **pH Paper.** A little more expensive, but infinitely more accurate, pH paper turns different colors according to the pH of the substance being tested. These colors match a chart that tells you the pH of that particular medium.
3. **Mechanical pH Meter.** Much more accurate than either form of paper, the pH meter is a great deal more expensive. It is generally found on small-scale commercially manufactured equipment and works in conjunction with microcircuits to automatically adjust the pH, primarily during mashing, and, to a lesser extent, during fermentation.

Adjusting pH. It does little good to know the pH of something if you cannot change it when you want to. Adding large quantities of water to either an acidic or alkaline solution will eventually raise or lower the pH, but it is not always feasible or desirable to increase the water volume. The easiest way to adjust pH is to add alkali to an acidic solution, and add acid to an alkaline solution. It is entirely possible to neutralize an acidic or alkaline solution by adding an alkali to the acidic or an acid to the alkaline to make a pH 2 a pH 7, or a pH 12 a pH 7.

OTHER TESTS

As you manufacture alcohol, there are certain stages at which it is desirable to know the status of your reactions. For instance: during hydrolysis, how much sugar have you made and/or when is hydrolysis complete; during fermentation, are you making acetic acid or alcohol; and what is the final proof yield of the alcohol produced? Yes, you can make alcohol without tests, but your waste factor is tremendously increased and efficiency enormously decreased.

Alcohol. Alcohol is measured (gauged) by proof, which is double the percent alcohol in solution. For example, a solution of 70 percent alcohol and 30 percent water is 140 proof alcohol; 90 percent alcohol, 10 percent water is 180 proof alcohol, and so on. The most common method for measuring proof is by using a hydrometer, of which there are two types: a low-proof hydrometer and a high-proof hydrometer.

A hydrometer works on the principles of specific gravity and density. Any object placed in a liquid will displace its own weight in liquid, whether it sinks or floats. If it floats, it will sink to different levels depending on the specific gravity of the fluids into which it is placed. The specific gravity of water is used as the base number for specific gravities, and is given a rating of one. Those fluids thicker (more dense) than water have higher specific gravities, and those that are thinner have lower ratings than one. The thinner the fluid, the deeper an object will sink into it. A hydrometer is graduated to show the differences in depth, i.e., difference in specific gravity between water and the medium being tested. The specific gravity of alcohol at 60°F is .794; of gasoline, .700–.780; and of water, 1.000.

A low-proof hydrometer is used primarily to measure the proof of your fermenting mash, which will generally not exceed 40 proof (20 percent alcohol by volume). A high-proof hydrometer is used to test the finished product and measures up to 200 proof. Hydrometers are not expensive—a couple of dollars—but you can make your own if you so desire. Use any hollow, glass cylinder (ideally 6" × ½" in diameter) that will float and can be weighted on one end (with BBs or lead splitshot) so it floats vertically. Place it in water, and mark the waterline (point to which the cylinder sinks) with indelible ink. Since gasoline is so close to the specific gravity of alcohol, you can use gasoline as the first constant. Place your cylinder in gasoline and mark the level to which it sinks. For the third constant, use ordinary rubbing alcohol, which is usually 140 proof, and mark the level to which the cylinder sinks. You now have three constants from which to work. Divide the distances between each of the three marks equidistantly so that by assigning 0 proof to the waterline and 200 proof to the gasoline line, you can determine the remaining proof lines. A hydrometer is the fastest and easiest way to determine proof.

An easy test for high proof is to burn some. A teaspoon of 196 proof alcohol burns for about 90 seconds in a small metal lid. A teaspoon of 140 proof alcohol burns for about 65 seconds. The burning time is directly proportional to the percent (or proof) alcohol present. Once a constant is developed on a known proof, simple arithmetic will put you within a few points of your product.

Still another simple method of determining alcohol proof is the weight by volume of alcohol compared to the weight of an equal volume of water. This works because of the specific gravity differences between alcohol and water. The formula is:

$$\text{Specific Gravity} = \frac{\text{Weight of Test Fluid}}{\text{Weight of Water}}$$

The volume used is inconsequential provided the same volume measure is used for both water and alcohol. Example: one gallon of water weighs 8.33 lbs; one gallon of 200 proof (anhydrous) alcohol weighs 6.6 lbs.; 6.6 divided by 8.33 = .79, the specific gravity of anhydrous alcohol.

Using two constants, the specific gravity of water (1.0) and the specific gravity of anhydrous alcohol (.79), the rule of thumb is: for each one percent increase in alcohol specific gravity above .79, the proof decreases 10 proof points subtracted from 200 proof. If the alcohol specific gravity is .89, the proof is 100. This requires an accurate weighing device, preferably one that measures in increments of ounces, not whole pounds.

Sugar. During hydrolysis, it is advantageous to know the status of your starch-to-sugar conversion so that you can save time and heat energy. A taste test will not work here, especially if you are using acid to assist hydrolysis. Commercial operations usually use a saccharimeter, which measures the degree or angle to which the density and consistency of the sugar present will refract light. This is an expensive piece of equipment, and not ideally suited for testing mashing sugar because of the variety of solids present. A saccharometer is a sugar hydrometer, and is considerably less expensive than a saccharimeter. However, during mashing and hydrolysis, suspended solids will affect the final reading. You must have a semiclear fluid–one in which nearly all solids have settled out–to check the sugar density during hydrolysis. This wastes time, and perhaps energy. While you are waiting for the solids to settle, you must continue to cook the mash. If you have reached your optimum sugar content (15 to 20 percent), this means wasted heat and fuel for unnecessary cooking.

A simple and easy sugar test that works regardless of the amount of solution-suspended solids is an iodine test. It is a simple chemical reaction that will tell you whether starch or sugar is present.

To prepare an iodine solution, dissolve 10 grams of potassium iodide in one pint of room-temperature water. To this add 6 grams of iodine crystals and stir until dissolved. Store in a bottle that does not admit sunlight. An alternate solution can be prepared by using one part of a two percent tincture-of-iodine solution to nine parts water, and store as above.

The iodine test: Draw a small sample from your cooking mash, and slowly add increasing amounts of the iodine solution until the mash solution turns blue, or dark red to purple, or light yellow to light red. An iodine solution will change the color of a starch solution to blue, and will change the color of a sugar solution light yellow to light red. A mash color-change of dark red to purple indicates partial starch-to-sugar conversion, or hydrolysis. When you obtain a light yellow or light red mash-sample color, hydrolysis is complete and you can stop cooking. To become accustomed to the colors, prepare solutions of household corn starch and sugar, one teaspoon of each in four ounces of water in separate containers, and run your tests.

Acetic Acid. Acetic acid (vinegar) is detrimental to your fermentation process, for it signals the beginning of the end of alcohol production. The test for acetic acid is extremely simple. Draw a sample from your fermentation tank, let the solids settle, and add a few drops of the solution to a small glass of warm milk. If the milk does not curdle–become spongy–your fermentation is proceeding properly. If the milk curdles, acetic acid is present, which means that alcohol produced will progressively diminish and acetic acid produced will progressively increase. (More on this in Chapter Eight: Fermentation.)

CHEMISTRY GUIDELINES

1. Store *all* chemicals out of reach of and access to children. Do not allow personal skin contact with acids and alkalies. Do not breathe dust or fumes.
2. Do not expect miracles on your first attempts. If a reaction does not work properly, retrace your steps. Try variables on smaller quantities until you discover the source of your problem.
3. Inorganic chemical reactions generally occur immediately and spontaneously. Organic chemical reactions usually take time–from one minute to several days. Have patience with organic reactions.
4. Heat affects chemical reactions. Each 10-degree increase in temperature doubles the reaction efficiency (halves the reaction time), until an optimum, maximum temperature is reached. At that point the reaction will either stop, reverse itself, or begin to make new compounds–not necessarily those you planned. If you want to experiment with heat to step up your reactions and decrease the time factor, do so in increments of 10 degrees until you reach an apparent optimum temperature. (This does not apply to yeast. Yeast has an absolute range of 60°F to 95°F.)
5. Do take your time in mixing and measuring, and do check things frequently during your first few efforts. Never assume that everything is going just fine. Some reactions are not immediately visible or apparent, such as the formation of acetic acid.

SOURCES OF CHEMICALS AND EQUIPMENT

Most of what you need can be obtained from a hardware store and/or a drugstore. Some can be obtained only from a chemical supply house and/or manufacturer.

A chemical supply house can generally supply you with all the chemicals and equipment you need–acids, alkalies, hydrometer, potassium iodide and iodine crystals, pH paper, etc. If there is no supply house in your area, do not overlook the possibility of a mail-order chemical supply house.

A hardware store usually sells muriatic acid (HCL) but does not sell sulfuric acid (H_2SO_4).

A drugstore will usually sell tincture of iodine; and may sell a hydrometer. If you cannot find a hydrometer, buy a urinometer. It too works on the principle of specific gravity.

A privately owned wine store (some states have only state-owned wine and liquor stores) may have the hydrometer, saccharometer, and pH paper you want.

5
Carbohydrates

CARBOHYDRATE means hydrate of carbon, or more simply put, water with carbon. Carbohydrates are organic compounds, including sugars, starches, and cellulose, with the general formula $C_x(H_2O)_y$. Glucose and fructose, simple sugars, follow this formula, for there are twice as many hydrogens as oxygens: $C_x(H_2O)_y = C_6H_{12}O_6 =$ glucose or fructose. But other substances classed as carbohydrates do not follow this formula. As it applies to fuel alcohol, carbohydrates will be more restrictively defined as dealing with saccharides, which, ironically, include sugars, starches, and cellulose.

There are four major forms of saccharides: mono-, di-, tri-, and poly-.

Monosaccharides. Monosaccharides are simple sugars that cannot be broken down—hydrolyzed—to a simpler form. The best-known are glucose and fructose. The ratio of carbon to hydrogen to oxygen atoms is 1:2:1 or $n(CH_2O)$ where *n* is between 3 and 8; therefore, monosaccharides go from $C_3H_6O_3$ to $C_8H_{16}O_8$. Each monosaccharide is composed of a carbon chain with an aldehyde or ketone functional group. If it has an aldehyde functional group, it is known as aldose. If it has a ketone functional group, it is called a ketose. Monosaccharides are also described by the number of carbon atoms they contain. For instance: triose (3), tetrose (4), pentose (5), hexose (6), etc. If a monosaccharide has carbons with an aldehyde functional group, it is known as an aldohexose; if it has 6 carbons with a ketone group, it is called a ketohexose. They look like this:

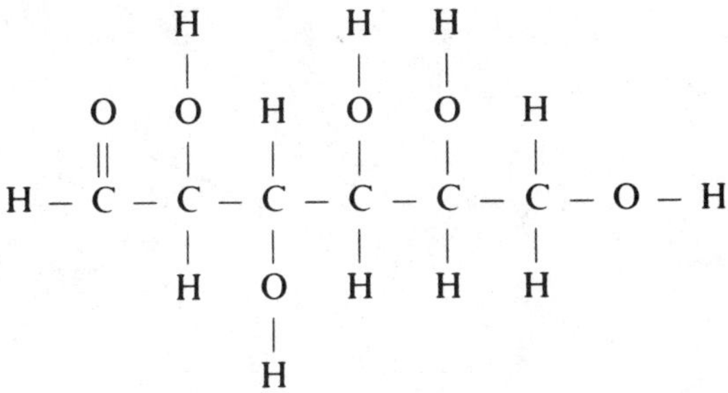

Glucose (Aldohexose)
$C_6H_{12}O_6$

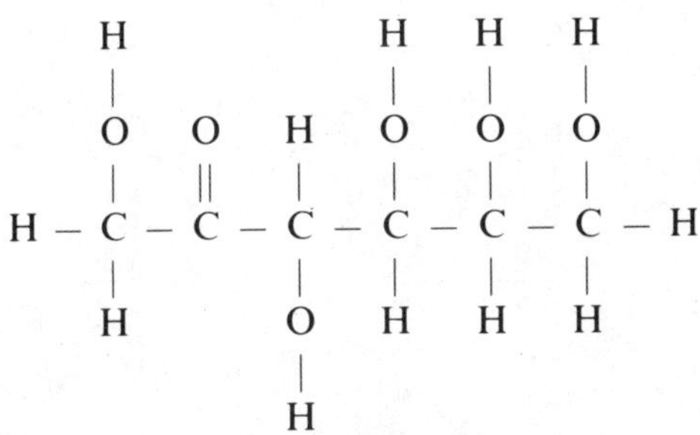

Fructose (Ketohexose)
$C_6H_{12}O_6$

Plain glucose is called D-glucose, or right-hand glucose. When all of the atoms are arranged exactly the opposite of D-glucose, it makes the mirror image of D-glucose, known as L-glucose, or left-hand glucose. The only difference between *D* and *L* is the direction each refracts a ray of light. The characteristics and behaviors are identical except for that.

Monosaccharides are straight-line chains, but in the presence of water (or within a living system) they tend to form rings. When two rings join, they form a disaccharide, meaning two sugars, and look like this:

Item One

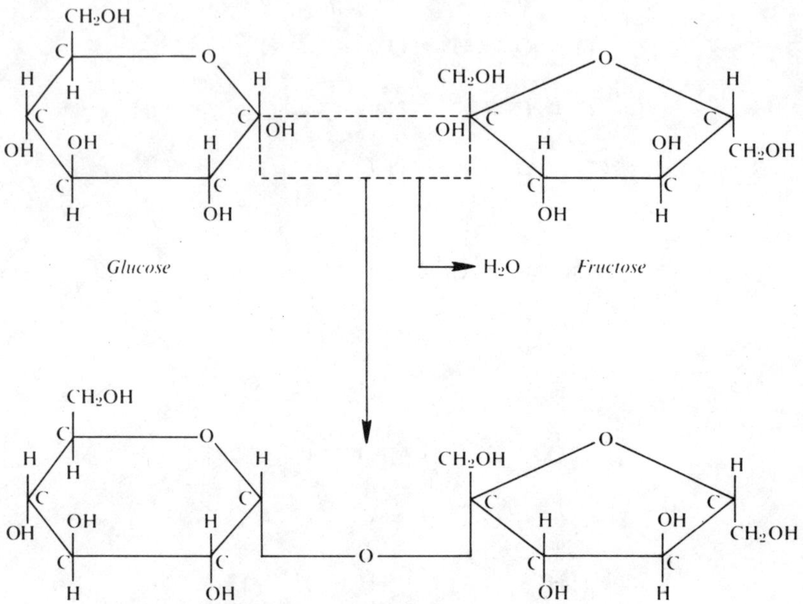

Sucrose
$C_{12}H_{22}O_{11}$

Disaccharides. When glucose and fructose combine, they form one molecule of water and one molecule of sucrose, a disaccharide. The dotted-line box between the glucose and fructose in Item One represents the bonding of the two sugars, and subsequent release of water. During hydrolysis, just the reverse happens—a water molecule cleaves the sucrose, sending an *H* to one side and an *OH* to the other, forming glucose and fructose.

Sucrose, obtained primarily from sugar beets and cane, is the most abundantly produced disaccharide in this country. Others include maltose, produced by the enzymatic action of diastase on starch, and lactose (milk sugar) produced from glucose and galactose and obtainable from whey. All must be broken down to simple sugars before fermentation can occur. However, in the absence of hydrolysis, enzymes, produced by yeast and other microorganisms, will break the disaccharides down to simple sugars.

Trisaccharides. Are formed in similar fashion to disaccharides, and when broken down, form three simple sugars. Raffinose is an example of a trisaccharide.

Polysaccharides. Starch and cellulose are both high-molecular-weight saccharides, consisting of long chains of glucose linked by a single-bond oxygen between each glucose molecule. Although starch and cellulose are extremely similar in chemical structure, their differences are quite apparent in real life. Cellulose is more complex than starch, with the addition of lignin to hold the cellulose fibers together. Nevertheless, both must be hydrolyzed to simple sugar before either is of any value for fermentation. As you will see in Chapter Seven: Mashing, the process is quite different for each. However, both do break down to glucose.

Starch is the principal storage form for sugar in plants and animals. In plants, starch occurs in two forms, amylose and amylopectin, which both consist of glucose units linked together. When the plant needs sugar (food) the starch is broken down into glucose units. In animal tissue, the starchlike substance is glycogen, which converts to sugars as the body needs energy. Glycogen means "sugar former."

There are straight-chain starches (those in potatoes) and branch-chain starches (those found in grains). However different they are, both will eventually hydrolyze to simple, fermentable sugars, as will cellulose.

SACCHARIDE PROFILE

Monosaccharides: Single Sugars.

Glucose (same as dextrose). The primary base unit for fermentation and the formation of starches, cellulose, and other saccharides. Occurs naturally in fruits and honey. It is also formed by the hydrolysis of higher saccharides.

Fructose. Generally combines with glucose to form disaccharides and higher saccharides. It also is found naturally in fruits and honey.

Disaccharides: Two Sugars.

Sucrose. A combination of glucose and fructose, sucrose is obtained primarily from sugar beets and sugar cane. Hydrolyzed, it yields simple, fermentable sugars.

Maltose. This is formed by the enzymatic action of diastase (the vegetable form of amylase). The enzyme maltase then breaks the maltose down to simple, fermentable sugars.

Lactose. Commonly called milk sugar, it is obtained by evaporating whey, which crystallizes into lactose. (*Note:* Lactose requires special yeast for fermentation.)

Trisaccharides: Three Sugars.

Raffinose. Primarily found in cotton seeds and sugar beets.

Polysaccharides: Complex Sugars.

Starch. Formed by long chains–either straight-line or branch–of glucose, it can be broken down to simple, fermentable sugar by either enzymatic action or hydrolysis (molecular water cleavage).

Cellulose. Formed by long branch chains of glucose, it is converted to simple sugar by the enzymatic action of cellulase, after the lignin bonds have been broken by chemical action.

Notice that, at first glance, starch and cellulose appear to be very, very similar; closer comparison reveals that they are extremely different.

RULES OF THE GAME

If a yeast will not ferment D-glucose, it will not ferment any other sugar. With few exceptions, if a given yeast will ferment lactose, it will not ferment maltose, and vice versa. All sugars must be broken down to the hexose level for fermentation to take place. This can be done enzymatically or hydrolytically.

Disaccharides are generally broken down to hexose level prior to passing through a yeast's cell wall for fermentation. However, it is possible for a disaccharide to enter the cell where internal enzymes reduce the disaccharide to hexose level for fermentation.

A Portion of a Starch Molecule

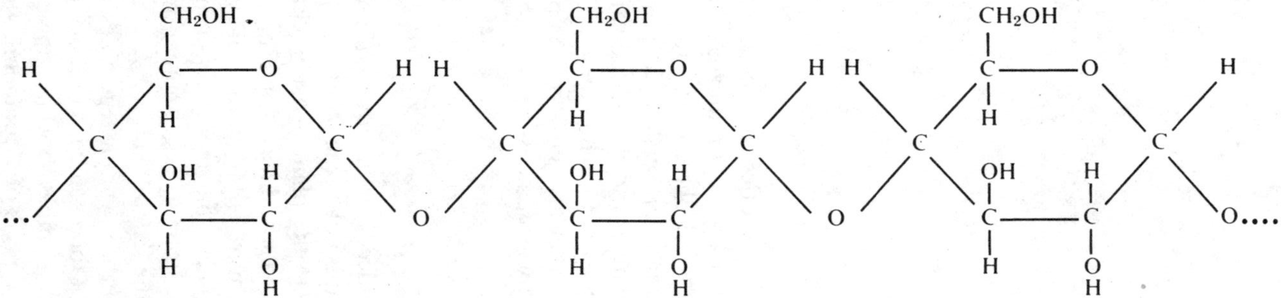

A Portion of a Cellulose Molecule

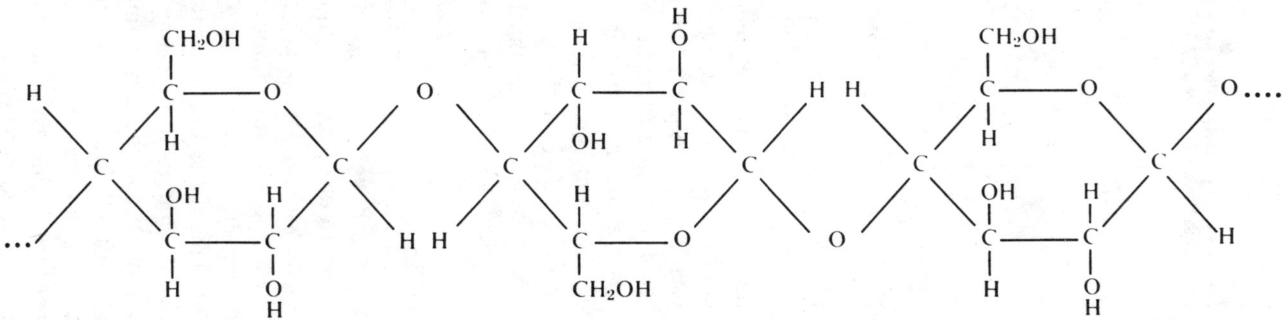

6
Microorganisms

MICROORGANISMS are microscopic, living entities of unicellular or multicellular form, including yeasts, molds, bacteria, protozoa, and viruses. As they apply to the manufacture of alcohol, yeast and the effect of mold and bacteria on the fermentation process are the primary concerns and interests here. It is important to understand the various functions and behaviors of mold and bacteria, for while they do have certain standards and ranges of behavior, they also show certain peculiarities that can spell the difference between success and failure in your alcohol production process.

YEAST

What Is Yeast? Yeast is a single-cell (unicellular) microorganism belonging to the plant kingdom, although it is not a true plant, i.e., yeast contains no chlorophyll. Yeast belongs to the plant division Thallophyta—those plants without true stems, roots, and leaves, and not producing flowers or seeds. There are over 1,300 strains of yeast throughout the world, so it is small wonder that there are so many misconceptions surrounding it; what one strain will do, another might not; what one strain is blamed for, another is guilty of.

Yeast is found in nature: on the skin of fruit, within vegetables, and even on flowers, where it uses nectar as a food source. You have perhaps seen evidence of yeast on the skins of grapes—a white, powdery-looking residue. It is a parasite with no powers of locomotion, which means it must attach itself to a host and live out its life as best it can. Although yeast will accept almost any nutrient molecularly small enough to be absorbed through its cell walls, its favorite nutrient (food) is sugar, found in fruits and rotten or rotting vegetables.

Yeast is best-known for its fermentative qualities in making alcohol, and as a leavening medium in making bread and other bakery products. However, some yeasts have low or no fermentative ability, and some (commonly called wild yeasts) are detrimental to the manufacture of alcohol and/or bread.

There are two basic types of yeast, informally called top yeast and bottom yeast. Top yeast does most of its fermenting near the top of the mash, and is a slower-acting yeast than bottom yeast. Bottom yeast more adequately permeates your entire mash solution and offers quicker and more thorough fermentation.

As far as fuel alcohol production is concerned, the yeast with which you are most concerned are strains from the genus *Saccharomyces,* species *cerevisiae* (baker's yeast) and *carlsbergensis* (brewer's yeast). Baker's yeast can be purchased at any food store in either dry or moist form. The dry is preferred for storage longevity; the moist must be refrigerated and has a relatively short storage life. Both dry and moist are similar in terms of use and preparation for fermentation purposes. Brewer's yeast is a top yeast and less desirable than baker's yeast, which is a bottom yeast.

Aside from fermentation, yeast is an ideal vitamin synthesizer. It has the ability to absorb and concentrate quantities of vitamins and by-products well suited as high-protein and high-nutrition food supplements.

THE THREE STAGES OF YEAST

There are three major stages or phases through which yeast passes.

1. Vegetative. The reproduction phase.
2. Fermentative. The anaerobic conversion of sugar to alcohol and CO_2 phase.
3. Respirative. The aerobic conversion of sugar to water and CO_2 phase.

Vegetative. Under a microscope, yeast cells have the appearance of fish eggs. Each cell contains a nucleus, which is where the vegetative process begins, surrounded by cytoplasm and a permeable, translucent membrane as the cell wall.

There are two basic yeast cell groups–haploid and diploid. Haploid contains the full and basic number of chromosomes normally found in the mature cell. Diploid contains twice the number of necessary chromosomes. Baker's yeast (*S. cerevisiae*) is diploid; therefore, it can reproduce by converting to haploid state, sporulating, and forming a bud, which attaches to the outside wall of the cell. This daughter cell remains attached until it reaches maturity, or, in some instances, until the mother cell generates another daughter cell, at which time the first daughter cell detaches and is on its own.

One cell of baker's yeast will generate from ten to forty daughter cells during its life. At the onset, the mother cell will generate a daughter cell every several hours. As the mother cell ages, daughter-cell generation slows down and reaches a point where reproduction takes from six to ten hours. Eventually the mother cell reaches its maximum production and dies. This will bear more significance in Chapter Eight: Fermentation.

Fermentative. During the fermentative cycle, yeast will convert glucose to two molecules of alcohol and two molecules of carbon dioxide using a glycolytic pathway under anaerobic (airless) conditions. Glycolysis is an energy-yielding metabolic breakdown of carbohydrates and sugars through a consecutive series of chemical and enzymatic reactions. Using a discriminatory selective process, the yeast cell membrane takes in simple hexose sugar molecules for food but excludes carbohydrates and other chemicals–regardless of their size. In the absence of suitable food of suitable size, the cell will send out enzymes to break down denser carbohydrate molecules into simple sugars that will fit through the membrane wall.

Once inside the cell, the sugar is phosphorylated (combined with phosphate) using ATP (Adenosine Triphosphate) present in the yeast cell, to form hexose diphosphate. Hexose diphosphate is then broken down through a series of oxidation and reduction reactions to yield carbon dioxide and acetaldehyde. The acetaldehyde is then reduced to alcohol.

At any point along the way, stray mold or bacteria can have an adverse effect on the fermentation process and the final products. Certain bacteria will create acids which may speed up or slow down reactions. Examples are: lactic acid bacteria (lactic acid), coliform bacteria (formic acid), and propionic acid bacteria (propionic acid). Acetobacter (acetic acid bacteria) will oxidize the final-step acetaldehyde and instead of alcohol, you will get acetic acid–vinegar. Mold will generally act as an inhibiting and spoiling mechanism.

Respirative. The majority of the time, the respirative cycle occurs under aerobic (oxygen-present) conditions, although yeast may anaerobically (without oxygen) respirate on occasion. During respiration, yeast continues to feed (produce energy and by-products), but the end products are water and carbon dioxide (by oxidation) instead of alcohol and carbon dioxide (by reduction).

Under aerobic conditions, mild fermentation yielding small traces of alcohol *does* occur, but respiration and vegetation dominate the yeast's aerobic activities. Contrary to popular belief, aerobic exposure does not automatically create acetic acid.

ADVERSE EFFECTS ON YEAST

Temperature. Yeast has a narrow temperature for effective activity range–about 60°F to 95°F. Yeast becomes dormant at about 60°F and below. While some cells will die at this temperature, the majority will simply hibernate. In fact, you can freeze yeast to -108°F (the temperature of dry ice) and subsequently revive it, at which time it will resume normal status and abilities (fermentation, vegetation, etc.).

The optimum temperature range for yeast's activities is room temperature–68°F to 75°F. During fermentation yeast will generate heat, as high as 85°F, and will not really start dying until about 94°F to 95°F. However, although yeast will ferment about the optimum temperature range, and survive right up to maximum, it is best to hold the temperature between 68°F and 75°F for best results.

Chemicals. Any chemical that can be used for food preservation will generally have an adverse effect on yeast and fermentation: sodium chloride, calcium or sodium propionate, sodium benzoate, and citric acids are a few. Phosphoric acid, salycylic acid (aspirin), sulfuric acid, and boric acid are detrimental to yeast, although their effects can be neutralized by the addition of a base (alkali) to adjust the pH factor. Water containing chlorine or fluoride will inhibit yeast and fermentation. Most chemicals used as antiseptics (such as hydrogen peroxide, iodine, and alcohol) will adversely affect yeast.

Osmotic Pressure. In solution, certain chemicals exert osmotic pressure on the yeast cell membrane and, if great enough, will rupture the cell wall. Sugar solutions in concentrations denser than 40 to 50 percent will kill yeast osmotically, as will 10 to 15 percent solutions of sodium chloride (table salt).

pH Factor. Although acids are generated during the fermentation process, excess acid is detrimental to yeast, as is an excess of alkali. Ideally, the pH of the yeast growth medium will range from pH 3 to pH 5, with an optimum pH of 4.5.

Alcohol. It is interesting that one of the products of fermentation will actually kill the yeast responsible for its production, or at least render it dormant. The average alcohol yield of fermentation is 10 to 11 percent, and up to that point there is little problem. Extended fermentation past 11 percent will eventually, under favorable conditions, produce a final yield of 13 to 14 percent alcohol. At that point, the yeast either dies or becomes dormant.

Ultraviolet Light. Ultraviolet light can be used to kill mold and bacteria, but in the process of eliminating these undesirables, you will kill the yeast as well. Ultraviolet light can be used during certain phases of alcohol production to eliminate mold and bacteria, but only when it is unimportant whether or not the yeast also dies.

MOLD AND BACTERIA

Mold. Unlike most yeasts, which are unicellular, mold is multicellular and forms mycelium. Mycelium is a group or cluster of cells attached at some point on their respective cell walls. Mold is asexual and reproduces by sporulation, and is generally recognized by a fuzzy or cottony look. Like yeast, mold is a member of the plant division Thallophyta, more accurately, of the subdivision Eumycetes, or true fungi. Unlike yeast, mold can grow at extremely low temperatures, and is pH-tolerant within a range of pH 2 to pH 8.5, although it favors an acid pH. Because of these characteristics, mold contamination of your alcohol production processes is not only likely but almost inevitable. Once contaminated, it is extremely difficult to get rid of, for that which kills mold also kills yeast.

Bacteria. Typically unicellular, bacteria can also be multicellular and form mycelium. Bacteria reproduce by cell division and the formation of spores. Bacteria have a much greater range of tolerances than either yeast or mold, so your alcohol production is quite susceptible to bacterial contamination in all areas of consideration: temperature, food substrate, and chemical composition.

Some strains of bacteria can withstand boiling water, whereas yeast and mold cannot. Certain species can survive pH changes during alcohol processing of alkali to acid or vice versa. Food and nutrients that one bacterium would reject, another will accept, so it is literally impossible to control all bacteria by formula changes. As mentioned previously, certain species not only affect chemical reactions but also instigate them. Once your operation is contaminated it is literally impossible to eliminate the problem without flushing your entire system and beginning again with fresh and noncontaminated materials.

Your greatest defense against mold and bacteria is preventive in nature. Cleanliness between operation processes is absolutely mandatory.

ENZYMES

Enzymes are protein catalysts that initiate, assist, or accelerate chemical reactions. In alcohol production the two major functions of enzymes are to hydrolyze starch and cellulose to sugar, and to assist in the fermentation process. Enzymes are generally characterized by the ending *-ase,* such as in amylase, cellulase, diastase, etc. They are found in nature, such as within the germinating seeds of grains, saliva, and within the cells of microorganisms, and are commercially prepared by culturing strains of select bacteria, primarily *Bacillus* and *Aspergillus* for starch hydrolysis, and *Trichoderma viride* for cellulose hydrolysis.

Hydrolysis. In order to obtain nutrition and energy, a living cell must come into contact with something small enough to penetrate the cell wall without damaging the wall. As polysaccharides, starch and cellulose are too large for cells to absorb and utilize as energy sources. They must therefore be broken down to simple sugars that will penetrate the cell wall, which means breaking the long chain of sugars that comprise the polysaccharides.

The most effective enzyme for starch hydrolysis is amylase in one form or another. Certain Indian tribes still chew kernels of grain prior to fermentation to hydrolyze the starch. However, the traditional and time-honored method of producing amylase is controlled grain sprouting. As grain sprouts, amylase (sometimes called diastase, the vegetable form of amylase) is generated within the germinating seed. The starch within the grain is not hydrolyzed because the shell of the germinating seed is so thick.

When sprouting is complete, the sprouted grain is then dried, milled (ground into meal), and called malt. When this malt is added to a starch slurry, the amylase hydrolyzes the starch. A slurry is a solution of suspended particles (starch or cellulose, etc.), generally about as thick as pea soup. (Details on sprouting in Chapter Seven: Mashing.)

Enzymes that work outside a cell are called exo-enzymes; those that work within the cell are called endo-enzymes. In a medium or substrate that contains no nutrition of permeable size, a cell will send out enzymes (exo) to break down whatever is available for subsequent admission to the cell for digestion. The general pathway is that a polysaccharide is split by an exo' alpha amylase into a dextrin. The dextrin is admitted into the cell where an endo' beta amylase reduces it to maltose, at which point maltase converts the maltose to glucose, which the cell utilizes by fermentation or respiration.

The most effective enzyme for cellulose hydrolysis is cellulase. The cellulase breaks down the cellulose fibers into fermentable sugar. Cellulase is made commercially by culturing the bacterium *Trichoderma viride.* The problem with cellulose hydrolysis is that it is a developed technology and not readily available to the back-yard enthusiast.

To hydrolyze cellulose, the lignin that binds the cellulose fibers together must be dissolved. This requires extended cooking (several hours) in either a sulfate or sulfite bath. A sulfate bath usually uses sodium hydroxide and sodium sulfide. The sulfite bath is generally composed of a bisulfite (sodium, calcium, etc.) and sulfuric acid. Unless the cellulose slurry of the bisulfite bath is washed thoroughly, it will present an eventual fermentation problem. The bisulfite will fix the glycolytic pathway at the acetaldehyde stage and instead of ethanol as the final product, you will get glycerol.

All is not lost, however, because of the complexity of the operation. Paper-mill cellulose wastes are ideal for cellulose hydrolysis, provided you have access to the material and adequate disposal facilities for your processing wastes. Be certain to determine the bath used for lignin separation. If it is sulfite, you will have a fixation problem during fermentation.

In lieu of sprouting grain to make amylase (which is time consuming) or letting nature take its course (also time consuming), you can buy commercially prepared enzymes, which are not only easier in practical application but also more expedient to your total operation. The only problem with commercial enzymes is acquisition. Most large industries that manufacture enzymes sell in large quantities—50- to 100-pound lots—with an average shelf life of 60 to 90 days. It takes only a few grams of these enzymes per bushel of grain used to make ethanol, and there are 454 grams in a pound. It is quite obvious that unless you have a large-scale operation, you will never use 50 pounds of enzyme within the 60- to 90-day life expectancy, and will end up throwing a good deal of it away. There are small businesses that buy the larger quantities and repackage them in smaller lots for sale to small-scale operations. Contact information on a few of these sources is given at the end of this book.

FUTURISTIC SUPER FERMENTERS

Although yeast has historically been the staple fermentation microorganism for alcohol production, it will, in all probability, be replaced by bacteria within the next decade. In 1980, a bacterium known as JW200 was discovered in Yellowstone National Park by a University of Georgia scientist. This bacterium reportedly digests organic matter fifteen times faster than yeast, has a higher alcohol yield, and uses more of the organic substrate than yeast does. The significance to alcohol production is quite obvious: higher yields in less time at lower cost. Other bacteria with as ideal if not better fermentation characteristics will surely be discovered or developed within the very near future, making alcohol production much more feasible and economical than it now is.

Unfortunately, these super fermenters are not yet available to the average small-scale alcohol operator; and if major alcohol manufacturers have their way, they *will* suppress the distribution of these microorganisms. Why? Market and product control. The easier and cheaper it is for small-scale operations to make their own, the larger the number of do-it-yourselfers who will take the steps toward fuel independence by making their own. Several hundred per state bothers no one. But several thousand per state who may eventually go medium to large scale is another matter—very much cause for concern to corporations currently investing millions of dollars in new-plant construction and existing-plant expansions. With that much invested, they do not look upon any form of competition with too much favor.

Current yeast and enzyme producers/distributors are adequately geared to mass-produce these bacteria as they are developed. They have the culturing facilities necessary to grow bacteria colonies, equipment necessary to isolate and purify select strains, and market capabilities for mass distribution. But will they want to and/or will they be able to? Or will these super strains be suppressed?

Suppression can take several forms. According to a recent Supreme Court decision, it is possible to patent a microorganism. The use of many bacteria can then be suppressed by "Patent Pending" under the ruse that they can only be released when the discoverer is certain the strain will not be stolen. This keeps it off the open market. However, it does not prevent private or secret deals with major corporations—to use it provided the strain is not released or disseminated to the general public, other corporations, etc. Conjecture? Yes. Possible? Yes. Difficult to prove and do something about? Yes. In this instance it is literally impossible for the private sector to do anything about suppression, unless collusion is suspected and an investigation instigated. You would begin by contacting your state attorney general, and appropriate state representatives.

Once something is patented, the patent rights can be sold, so suppression can occur by having the patent bought and taken off of the market. This may someday conceivably happen, and may fall under antitrust and/or monopoly laws. It is again extremely difficult for the private sector to attack and rectify, let alone identify.

Bacteria producers/distributors may have a hand in suppression by mandating minimum-purchase requirements—volumes so high that the average person could not possibly afford to buy the bulk quantity. This will perhaps be the easiest suppression to offset. By simply forming co-op purchases, a group of small-scale operators can band together and make one purchase to be divided among them. In this instance, the manufacturer must sell to you at the same price he sells to corporations provided you satisfy the minimum bulk requirements. The considerations here are: Will the price justify the final product, and is the bacterium that much better than yeast?

7 Mashing

MASH IS THE feedstock used for fermentation and most anyone who can boil water and measure—a cup of this and a pound of that—can make it. Well, it's almost that simple. There are a few do's and don'ts along the way, and a lot that *can* go wrong, but not too many problems that cannot be avoided or handled.

Regardless of the feedstock you choose—corn, potatoes, sugar beets, etc.—cleanliness throughout the entire process of mashing and fermenting is imperative. True, you can make several consecutive batches of mash, ferment them, and simply empty the tanks in readiness for the next batch, but sooner or later mold and/or bacteria will catch up with you and ruin an entire batch. Residual acetic acid may also build up to the point of kicking off a batch of vinegar instead of alcohol. If it happens, it will cost you raw material and time consumed, and leave you with one big mess to dispose of. Before the use of fresh equipment, and after each use of equipment, wash, flush, and soak everything with clean water and a weak bleach solution, or sodium hydroxide.

The equipment for mashing is basic and simple—mostly containers for soaking grain, and metal utensils (pots) for cooking.

THE BASIC MASHING PROCESSES FOR GRAIN, SUGAR BEETS, SUGAR CANE, AND POTATOES

Mashing is converting your raw material(s) into a fermentable feedstock. Although the basics of fermentation are about the same—using yeast to ferment glucose into alcohol—there are dissimilarities among various feedstocks in preparing mash. Grain and potato starches must be hydrolyzed to sugar before fermentation, whereas sugar beets and cane directly render sugar during mechanical processing.

STEP-BY-STEP GRAIN MASHING

Although any grain can be used, corn will be used for the example since it is the most popular and widely used grain feedstock. Sprouting grain to be used for starch hydrolysis is the first step in grain mashing. However, if you choose to use commercially prepared enzymes (amylase) for starch hydrolysis, you can skip sprouting and proceed to the second step.

The basic grain mash formula for each three gallons of ethanol desired is eight parts cooking grain, and one part sprouting grain which will subsequently be added to the cooking grain when it is cooked (hydrolyzed). If you choose to use enzymes instead of sprouted grain, the process is given on page 77. To this basic formula is added four ounces of yeast. (See Yeast Inoculation: page 87.)

Grain. The basic grain mashing process is:

1. Soaking dry grains for two to three days.
2. Sprouting grain (malting) to be used during grain hydrolysis.
3. Drying sprouts (malt).
4. Gristing grain (making ground malt).
5. Cooking grain to hydrolyze starch to sugar (using sprouted grain or enzymes).
6. Combining materials and ingredients at a fixed temperature.
7. Setting up mash for fermentation, and final fermentation.

Sprouting. The first step of corn mashing is sprouting corn to make corn malt. Fresh corn, on or off the cob, will not work. You must start with dried whole-kernel corn off the cob, just as you would use a dry seed to plant and grow a stalk of corn.

1. Begin with a minimum of 7 pounds of dry corn for every bushel of corn you plan to mash, so that you have an eight-to-one ratio of mashing corn (eight parts) to sprouting corn (one part). Of course you can sprout more than needed and store it. Fill any container no more than half full with the dry corn. This allows plenty of room for water absorption and corn swelling. The corn will just about double in weight and volume when hydrated.

(Five- to ten-gallon plastic buckets are ideal for this step.) Add cold tap, stream, or well water to double the volume of the corn; i.e., if the corn level is one fourth the container's volume, add water to the halfway mark. Stir the corn for a few seconds, and then use a sieve or a piece of plastic screen to scoop off floating debris. Anything that floats will not sprout. Let this sit at 50°F to 60°F—more or less—for 24 hours.

NOTE: Caution is advised if you use stream or river water. Be certain that the water is not polluted and, above all else, does not contain polychlorinated biphenyls (PCBs). PCBs will contaminate your mash and resultant DDG, which will carry over to the livestock that eats it, and to the subsequent consumer of the livestock meat or products (milk and eggs). PCBs are reputed to be carcinogenic, so if there is any doubt whatsoever about the quality of your stream or river water, check with your nearest EPA office. If commercial enterprises have access to the river or stream, the odds are good that there is some form of contaminant or pollution in the water. If you would not personally drink the water, do not use it for mashing and fermentation.

2. After 24 hours, pour off all water and refill with fresh water. Agitate the corn with either a wooden paddle or your hand; drain the water again and refill with fresh water. Let this sit for another 24 hours.
3. Repeat Step Two at the end of the 24 hours; refill with water and let it sit for a final 24 hours. This three-day process hydrates the corn in readiness for sprouting.

NOTE: You must drain the water each day, and stir the corn to prevent rotting and/or mold formation. Rotten corn will ferment, but it will not sprout. If by any chance you lose a batch of sprouting corn to mold, bacteria, or rot, save it for fermenting, and start a fresh batch of sprouting corn.

During the three-day soaking period, if you do not yet have a sprout box, you should build one. A sprout box can be either plastic or wood (avoid metal because of rust and metal ion contamination); a shallow 3′ × 4′ × 2″ is generally sufficient for each 7 pounds of grain you wish to sprout.

Begin with a comfortable base of ⅜-inch to ½-inch plywood cut to dimensions 3′ × 4′. Cut four 2-inch strips to length for the two sides and ends. Use white shop glue and finishing nails to attach the strips to the perimeter of the base. (The glue is not absolutely necessary, but it does offer a bit more durability to the box than simply nailing the pieces together.) The resultant shallow box must be lined with plastic both bottom and sides, since sprouting will bring moisture into constant contact with the bottom and part of the sides of the box. Use a single piece of plastic; do not try to cut, match, and glue. This means that you should "Christmas wrap" the corners, but you must preclude any moisture contact with the wood. Lap the plastic lining over the top edge of the sides and ends, and fasten with a row of staples all around the perimeter of the box. Cut off any excess plastic and your sprout box is almost ready to receive the hydrated grain.

Whether you use a prefab plastic box or homemade wooden box, cut a piece of cloth (flannel, wool, or cotton) to fit the bottom dimension of your box. The cloth should be colorfast, preferably no prints or patterns. Wash the cloth in hot water with bleach and rinse thoroughly. Wring most of the water from the cloth and spread it flat and even on the bottom of your sprout box. The cloth must be damp–not too wet and not too dry–for successful sprouting.

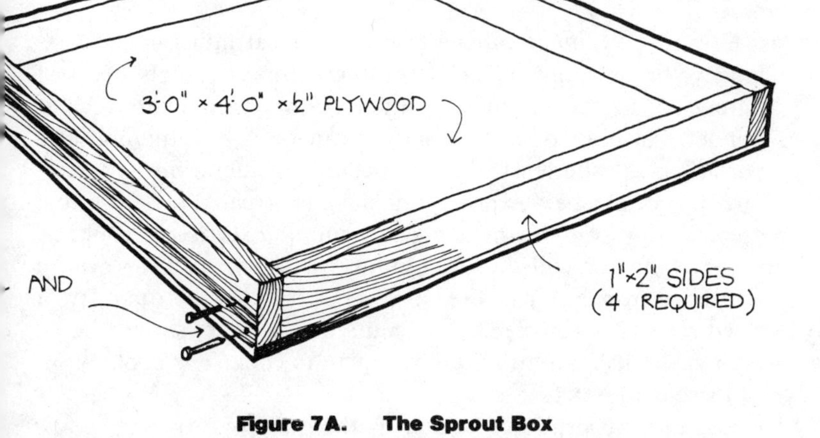

Figure 7A. The Sprout Box

4. At the end of 3 days, drain the water from your soaking grain, and spread the corn in a thin layer—½ inch maximum—on top of the damp cloth. Cover the box with a transparent or translucent plastic sheet to retain moisture and admit light. The box should be placed where it will receive indirect sunlight each day to diminish mold growth potential. Avoid direct sunlight, either indoor or outdoor, for it will have a tendency to cook the corn grain. Box temperature should be about 65°F to 75°F.
5. Each day for the next 3 to 5 days, check the cloth for dampness and add water as necessary. Also, gently roll the corn around each day—bottom up and top to bottom—to aerate the corn and help prevent mold formation. Depending upon temperature, the corn should begin to sprout on the second or third day. When the sprout is about 1½ inches long, it is malt and can be allowed to dry. If mold begins to develop at any time during sprouting, expose the corn to direct sunlight for 1 hour, stirring it around gently to expose all of the corn to the mold-killing sunlight. Then, check the cloth for dampness, add water if necessary, and replace the plastic cover.
6. Remove the plastic cover and cloth, and let the corn dry at room temperature or in the sun. If you dry indoors at room temperature, rotate the corn for the first day or two to prevent mold formation. Oven-drying at any temperature is the most expedient drying method, but requires energy, which costs money. When dry, the malt can either be stored whole kernel, or immediately milled (ground) and used for Step Two of the mashing process.

   Grinding the malt can be done at a local mill—which costs money—or by hand, using a meat grinder set on its coarsest setting. A motorized meat grinder is preferred over a hand grinder. An hour of hand grinding can be very fatiguing, and boring, to say the least. The important consideration is to rupture the grain's shell, exposing the amylase created by sprouting.

The second step of grain mashing is a simple exercise in cooking. Any pot or container will do—aluminum, cast iron, steel, or even a 55-gallon drum over an open fire—as long as it holds the quantity of hydrolyzed corn that you need. The container capacity should be 40 gallons for each bushel of dry grain you plan to cook. The cooking involves a three-step process:

1. Hydration. Absorption of water by the grain (corn).

2. Liquefaction. Conversion of insoluble carbohydrates to soluble carbohydrates.
3. Hydrolysis. Conversion of starch to sugar, using heat and enzymes.

Hydration. It is not necessary to soak cracked or ground corn prior to cooking, since boiling for 10 to 15 minutes both hydrates it and begins liquefaction. However, it is advisable to soak whole-kernel corn for 20 to 30 hours prior to cooking, for it would otherwise take about an hour of boiling just to hydrate and rupture the shell.

Although you can use whole kernel grain for mashing, the ideal cooking and hydrolysis condition is maximum exposure of starch granules/particles to water. This means that the more finely ground the grain, the faster and more complete the hydrolysis will be. With whole kernel grain, the starch particles are packed tightly together. Hydrolysis takes longer and is less efficient with whole kernel. The major consideration of grain status is the cost difference between whole kernel and cracked or ground grain compared against the extra cooking time and fuel cost for whole kernel hydrolysis.

Liquefaction. If you cook anything long enough, it will eventually gelatinize and thicken in solution (turn to mush). This is fundamental starch separation, and is necessary for any sort of efficient hydrolysis. The starch cells must be exposed to the amylase for conversion, and the more separated the cells are, the more rapid and efficient the hydrolysis. The addition of enzymes during cooking accelerates liquefaction and hydrolysis.

Hydrolysis. Hydrolysis is the final step and the major reason for cooking at all—to convert starch to sugar for fermentation. Basically, hydrolysis occurs when a water molecule cleaves a starch molecule in half (or breaks in at some point along a starch chain), resulting in reduction to sugar. As a complete starch chain breaks down, more and more sugar is created, depending upon where the water cleaves the starch. If the water molecule cleaves the very end of the chain, a sugar molecule is released. If it cleaves the middle, it creates two minichains, which are still starch but very susceptible to further cleavage.

Partial hydrolysis can be accomplished by simply boiling the starch. More complete and accelerated hydrolysis is obtained by the addition of enzymes. Introducing acid—either diluted or fuming hydrochloric or sulfuric–to the solution further enhances and accelerates hydrolysis. Acid should be gradually added until the pH factor is 2.5 to 3.5.

Cooking One Bushel of Grain. Whether your feedstock is whole kernel, cracked, or ground corn, use 20 gallons of water for each bushel of dry feedstock for the initial cooking. This will give you a total volume of about 26.7 gallons, for the corn will displace its own weight (56 lbs/bushel = 6.656 gals/bushel) in water.

The rule for water quantity is to use a minimum amount for cooking (no less than 20 gallons per dry-grain bushel), and a maximum amount for fermentation (about 37 to 40 gallons). The more water you use for cooking, the more fuel you will waste bringing everything to a boil. For fermentation, the less water you use, the less alcohol you will get; and if you use too much water you will waste fuel during distillation heating a larger mass. Since the average alcohol yield per bushel of grain is 2.5 to 3.5 gallons, and fermentation generally stops at 10 to 12 percent alcohol by volume, your optimum water quantity for fermentation is 37 to 40 gallons.

1. Place corn and water in cooking vessel and bring to a full boil. Let this boil for 15 minutes (45 minutes for presoaked whole-kernel corn), stirring frequently.
2. Add 2 to 3 ounces of baker's or brewer's yeast and continue to boil for another 15 minutes. The yeast will die within seconds, but the enzymes it contains will begin to work. Stir frequently, or starch lumps will form.
3. Add 3 pounds of ground malt and simmer (mildly boil) for 1 hour, stirring occasionally. The balance of your malt will be used during fermentation.

This process and these time limits will provide partial hydrolysis of starch to sugar, which means the total fermentation time will be just about double, or about five to seven days. Commercial hydrolysis requires double the above cooking times, which means double the energy input. But, the commercial operation gains the advantage of a lesser fermentation time, about two to three days, and commercial operations are interested in just that–how much can we produce in how short a period of time. For the small-scale operation, a longer fermentation period is of little consequence and is generally acceptable. If, for whatever reason, you prefer a shorter fermentation period, double the above time frames.

If you live in a mountainous or high altitude region, your cooking times must be increased slightly. You are probably already accustomed to the effects of lower atmospheric pressure on your regular home cooking—water comes to a boil faster and at a lower temperature. The exact cooking time variance will really depend upon your exact altitude, and will take a little trial and error to find the time best suited for you. A consolation to this extended cooking time for hydrolysis is that you will save some fuel energy when you distill your alcohol, because the vaporization temperature for alcohol (normally 173°F at sea level) will be lower, and complete distillation will occur faster.

Tips and Options.

1. When possible, cover the cooking vessel with a lid. This brings the solution to a boil quicker, and retains heat during cooking.
2. It is imperative that the grain not scorch while cooking. When stirring, be certain to scrape the bottom of the vessel thoroughly.
3. As you cook, the water level will drop because of steam escaping. Replace this water as necessary to maintain the original fluid level.
4. You may choose to use commercially prepared enzymes instead of malt. Depending on current market conditions and your source, these enzymes will add about 20 cents per gallon to the total cost of your ethanol. Stricter time tables and temperatures must also be observed when using commercial enzymes.
5. Acid may be used to assist in hydrolysis. It is introduced during Step Three 15 minutes after you add the malt. Diluted hydrochloric or sulfuric acid is most preferred. Add enough to bring the pH factor to pH 2.5 to 3.5. Extreme caution is advised in handling, storing, and mixing. Do not breathe the vapors—they can cause permanent lung damage and/or chemical pneumonia. Do not use citric acid, acetic acid, or phosphoric acid.
6. Once cooked, the grain and broth must be cooled to room temperature before the fermentation steps can begin. This can be facilitated by adding cold water to bring your total volume up to about 37 gallons per bushel of dry grain.

POTATO HYDROLYSIS

Potato hydrolysis follows much the same procedure and formula as grain hydrolysis with few exceptions.

1. Potatoes are generally shredded or pulverized prior to cooking; a simple operation since potatoes are about 80 percent water and easily worked.
2. Potatoes cannot be successfully sprouted to make malt. The potato will sprout *eventually,* but the amount of amylase produced is insignificant for practical use. The dark and damp conditions necessary for any sort of sprouting also encourage mold growth. If potatoes are used as feedstock, either grain malt or commercial enzymes must be used for hydrolysis and fermentation.

SUGAR BEET EXTRACTION PROCESS

Fermentable beet sugar is extracted by diffusion—as the sliced or crushed beets are cooked, the beet cells release their sugar into the surrounding medium. Since sugar beets contain from 12 to 15 percent sugar, you will want to add 7 times the volume of water to the wet weight of beets so that the resultant juice is about 20 percent sugar by volume. For each bushel of beets, then, add from 40 to 50 gallons of water.

1. Slice the beets wafer thin, or crush the beets to a pulpy mass.
2. Add the necessary volume of water and bring to a boil.
3. Cooking time will vary depending on the thickness of the slices or the degree to which the cells are exposed by crushing. This can be anywhere from 30 to 45 minutes. A saccharometer can give you the sugar percentage; or use an iodine test.
4. Once cooked, this slurry can be fermented as is, or the beet pulp can be filtered or pressed to separate the pulp and sugar juices.

SUGAR CANE EXTRACTION

Sugar cane extraction requires equipment that can be rather expensive and elaborate. Cane is the primary feedstock in Brazil, because cane is the primary crop in Brazil. However, in the United States, it is aggressively grown in only three southern states, and Hawaii. If you have ready access to sugar cane and the price is right, and if you can afford the equipment, the rest is easy.

1. Crush the cane stalk, and roller-press to squeeze and extract the juice (use a machine similar to the old-style wringer-washers).
2. Boil the resultant juice in any open container for 20 to 30 minutes to eliminate any stray microorganisms.

3. The juice will be slightly acidic, so it must be pH adjusted prior to fermentation. Once adjusted, it can be directly fermented.

DOING IT THE OLD-FASHIONED WAY

Moonshiner's Likker. If all of the above seems too much trouble, you can do as the old-timers used to do and make natural alcohol. Known by moonshiners as pure corn "likker," it utilizes only natural enzymes for mashing and only natural yeast and enzymes for fermentation. However, with the exception of sprouting-grain method, the time factor–start to finish–is three to four times longer. Instead of five to seven days, it takes twenty-one to twenty-eight days to mash and ferment a batch of "likker."

The basic formula is 3 parts corn to 1 part malt. The process is:

1. For each total bushel of corn and malt combined (3 parts to 1 part) add 37 gallons of water.
2. Cook either cracked or ground corn (do not use whole kernel) without malt for 30 minutes, and let it sit overnight (24 hours) in an open container.
3. The next day, add the malt and stir thoroughly.
4. Place this resultant mash in an anaerobic fermentation tank (see Chapter Eight: Fermentation) and leave it alone for the next 21 days. Addition of yeast is optional.

The enzymes involved are amylase/diastase (malt), and zymase (from the natural yeast in the cooked corn, now dead from cooking). A small amount of yeast is provided by the uncooked malt, but it is miniscule in quantity.

Potato Likker. The same basic process can be used with potatoes, although the quantities are somewhat different. Use a 5-to-1 ratio of potato to grain malt. Use 37 gallons of water per bushel of potatoes and cook the pulverized potato, including the skin, for 1 hour. Cool to room temperature and add the malt. Ferment anaerobically.

Applejack. More commonly called hard cider, applejack can be made and fermented into ethanol by using fresh or rotten apples, cooked or raw. There is enough sugar in apples so that all that is really necessary is to squeeze the juice from the apple and ferment the juice. No malt is necessary, and you may or may not–at your discretion–add yeast to the fermenting juice.

WHAT TO DO IF YOU HAVE MASHING PROBLEMS

1. Sprouting temperature. Any time the ambient temperature drops below 60°F, it discourages sprouting and encourages mold and bacterial growth.
2. Humidity will have a minor effect on drying sprouts, but more especially on storing grain and malt. Avoid damp areas.
3. If you are obtaining inadequate liquefaction or hydrolysis, you may need to extend the cooking time.
4. If lumps form during cooking, reduce or remove the heat, and stir the mash slurry until the lumps are gone. Then resume cooking.
5. Scorched or burned grain will not ferment. It is both a waste of raw material and money to let part of a batch scorch or burn, so stir frequently. However, if scorching does occur, it will have little or no effect on the unscorched mash and fermentation. Leave the scorched material in and proceed with fermentation.
6. Pressure cooking. A simple pressure cooker can be fabricated from a fifty-five-gallon steel sealable drum to which has been welded a salvaged pressure-cooker vent pipe that will receive a pressure regulator (the weight that jiggles and allows some steam to escape). The advantage of a pressure cooker is that your total cooking time is reduced, which means less energy input to cook a batch of mash. The disadvantage is that under pressure, the temperature inside the cooker will be much higher, depending on the exact weight of the regulator, so the possibility of scorching is that much greater. You cannot stir the mash in the sealed container, unless you have a mechanical stirring device built into your apparatus. While this is not impossible, it does require a little initiative and ingenuity to build. If you consider buying a commercial large-volume pressure cooker with a built-in agitator, you must consider the cost of the equipment compared to the energy savings over the life of the equipment. If your operation is too small to make it cost feasible, forget the commercial pressure cooker. Either build your own, or cook in an open container with a lid.

8
Fermentation

By its truest definition, fermentation is a metabolic, energy-yielding, oxidation/reduction reaction in which organic and inorganic compounds serve as hydrogen donors and acceptors. Fermentation, as far as you are concerned, is simply the conversion of sugar into alcohol using the mash you made in Chapter Seven, plus yeast and enzymes, under anaerobic conditions. And to do this, you need a fermentation tank.

THE FERMENTATION TANK: WHAT TO CONSIDER

Material. Almost any container that will hold water and can be sealed airtight can be used as a fermentation tank. The volume of your tank depends on your personal operation, so you can realistically work with anything from five-gallon plastic buckets with sealing lids to five-hundred-gallon vats with airtight covers. The tanks can be made from a variety of materials, although there are certain advantages and disadvantages to using plastic, glass, metal, or wood.

1. Plastic. Plastic is the most favored tank material because it cleans easily, it does not allow air to bleed through the sides, and it does not readily absorb odors and chemicals although repeated and continued use will eventually cause trace amounts to be absorbed. Plastic is also a slow conductor of outside heat. This property can become a detriment, however, because plastic does not allow heat generated by the fermentation process to

dissipate quickly through the plastic walls. Plastic fermentation tanks are therefore recommended only up to about a fifty-gallon capacity, unless a heat exchanger is incorporated into the design.

2. Glass. Glass has the same basic advantages as plastic, although glass absorbs fewer odors and chemicals during prolonged use. Glass is difficult to fabricate and, once constructed, extremely prone to breakage. If a glass tank is used outdoors, and exposed to direct sunlight, the heat generated will kill the yeast.
3. Metal. Metal is ideal for capacities over fifty gallons, for it allows the internal heat to dissipate more readily than do plastic, glass, or wood tanks. Iron, steel, and aluminum will contaminate your mash over a period of time, especially if you are making successive runs using the same mash feedstock. If used outdoors, a metal tank must be kept shaded from direct sunlight. Metal will become hot if exposed to the sun and transfer yeast-killing heat into the tank. Metal is subject to corrosion and rusting (iron and steel) or pitting (aluminum), unless electroplated with copper, or treated with an alcohol-resistant, nonreactive paint or coating. Stainless steel or copper are the most desirable metal tanks, but the high cost is somewhat prohibitive weighed against their advantages. It is best to invest in a good stainless steel or copper still if you really want to spend the money.
4. Wood. Wood is extremely favored because of its low cost. It can be easily fabricated, and mash contamination is virtually nonexistent. Most moonshiners usc wooden tanks for the above reasons, and also because the materials can be easily carried, piece by piece, into the woods. Because wood is porous, it must be sealed to prevent mash aeration (aerobic fermentation) and chemical absorption on the inside. The outside can be sealed with two coats of any varnish. The inside must be coated with an alcohol-resistant, nonreactive coating. Two coats of alcohol-base paint will work.

Regardless of the material that you choose for construction, there are four criteria that apply: capacity in gallons or cubic feet, ease of loading the tank, ease of emptying the tank, and airtight and watertight seal.

Capacity. The capacity depends upon your desired end-production volume. You will need a fermentation tank capacity of 16 gallons for each gallon of alcohol you want to produce on a single run—15 gallons for mash and 1 gallon dead-air space. As a rule of thumb, allow 40 gallons for each bushel of grain you plan to mash. At 7.48 gallons per cubic foot, this translates to roughly 5.5 cubic feet per bushel.

It is best to build a tank larger than you really need. This will facilitate your current fermentation plans and allow for expansion in the future. If you need ten gallons of ethanol per week, build a tank with a twenty-gallon capacity, which would be a four-foot-cube tank (sixty-four cubic feet). This can be easily constructed from three 4′ × 8′ sheets of plywood, of either 3/8-inch or 1/2-inch thickness.

Ease of Loading. Ease of loading (charging the tank) is best accomplished with a hinged lid. You must be able to dump your mash in easily and quickly with no waste or mess. Forget using a funnel; it is too slow and will clog. A lid also makes cleaning the tank after a run that much easier; you will be able to hose out and scrub down every square inch of the tank's interior.

Ease of Emptying. Discharging the tank is done in one of two ways. If you use a grain mash, you will have thick, undissolved feedstock settlement on the bottom of the tank, and semiclear liquid on top of that. To avoid excessive and time-consuming filtration, it is best to have two spigots—one for the liquid and one for the feedstock. For watery-thin beet or cane feedstocks, one spigot is sufficient. (See page 85 for spigot location and installation.)

Airtight/Watertight. Leaks mean loss of product—so your tank must be absolutely watertight. This can be done by placing strips of neoprene or rubber between seams, or by using butyl caulking or silicone gel during construction. The lid can be made similarly airtight. (See page 83 for construction details.)

Tank Location. The considerations here are: convenience, process efficiency or ease, temperature, smell, and noise.

Convenience is more a matter of space than anything else. Where will the fermentation tank(s) fit best—basement, garage, spare room, or outdoors? If possible, the tank should be indoors in the winter.

Processing will be easier and more efficient if the tank is located on a higher plane than the still. If you can use gravity to flow the fermented mash right into the still, then all the better. This is a minor consideration on small-scale fermentation, but on a five-hundred-gallon run, it is both tedious and time-consuming to transfer the mash by hand.

Temperature is critical in two ways—too hot and it will kill the yeast; too cold and it will inhibit fermentation or bring it to a standstill. More on this later under seasonal considerations.

Fermenting mash will have a distinct odor that some consider agreeable and others consider offensive.The smell is caused by esters, aldehydes, and other volatile products of yeast metabolism. If you have a sensitive nose, ferment a one-gallon batch of mash to test your tolerance before you decide on an indoor tank location. Family members must also be considered on this one.

As carbon dioxide escapes during fermentation, it makes a loud bubbling noise which, in the dead of night, can be somewhat unsettling. You may become accustomed to the sound eventually, just as people become accustomed to a dripping faucet or to living next to railroad tracks. Then again, it may eventually drive you right up a wall. Give this some thought.

HOW TO BUILD A FERMENTATION TANK

Wooden Tank. Begin by cutting three sheets of 4′ × 8′ × ½″ plywood (preferably marine grade) in half to create six pieces of 4′ × 4′ × ½″. Cut ½ inch from one end of two of these pieces for a final dimension of 4′ × 3′11½″ × ½″. These two pieces will be insert side panels.

Paint all flat surfaces (both sides) with two coats of alcohol-resistant paint or primer. Put two coats of varnish on all edges, full perimeter.

Use one piece of 4′ × 4′ as a base and attach two 4′ × 4′ pieces to the outside of the base facing each other. Use a sealant (gasket or caulking) at the joints; and fasten with ¾-inch to 1-inch wood screws—at least five on each side. (Drill pilot holes to avoid splitting the edge of the base.) (See figure 8A.)

Lay this U-shaped structure on its side and insert one side panel (4′ × 3′ 11½″). Seal and fasten down with wood screws to the bottom first; then attach to other sides.

Flip the unit 180 degrees and install the other side panel in the same way. (See Figure 8B.)

Trim the last piece for the lid so that it is flush with the outside

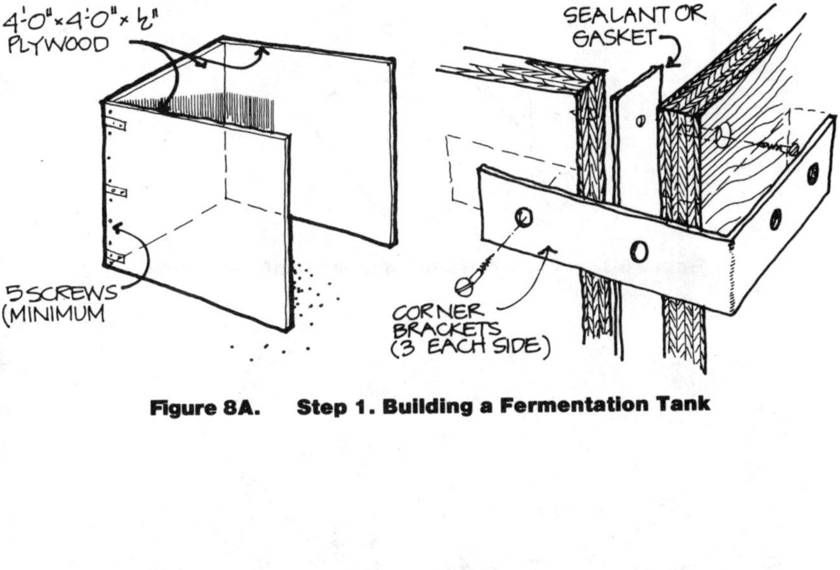

Figure 8A. Step 1. Building a Fermentation Tank

perimeter of the box, and attach with two hinges (or use a piano hinge). Tack a strip of neoprene or rubber to the top edge of the box, full perimeter, and a similar piece to the underside edge of the lid, full perimeter, so that they meet when the lid is closed. A small latch to secure the lid is optional. (See Figure 8C.)

Outlets. The thin-liquid outlet should be no less than ½ inch in diameter. For beet or cane fermentation, this hole can be drilled through any of the sides, exactly flush with the inside base bottom, i.e., ½-inch up from the outside bottom of the tank. Sugar cane or beet fermentation uses the juice extracted by diffusion or squeezing, so there is very little settlement on the bottom of the tank. For grain mash, the hole should be about 1 inch above the feedstock settlement level. This level

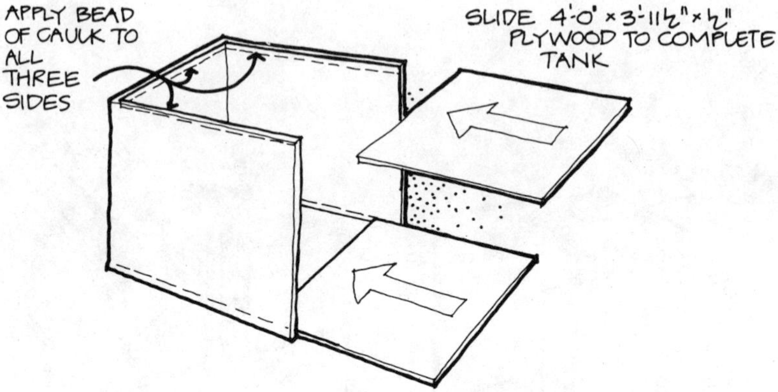

Figure 8B. Step 2. Building a Fermentation Tank

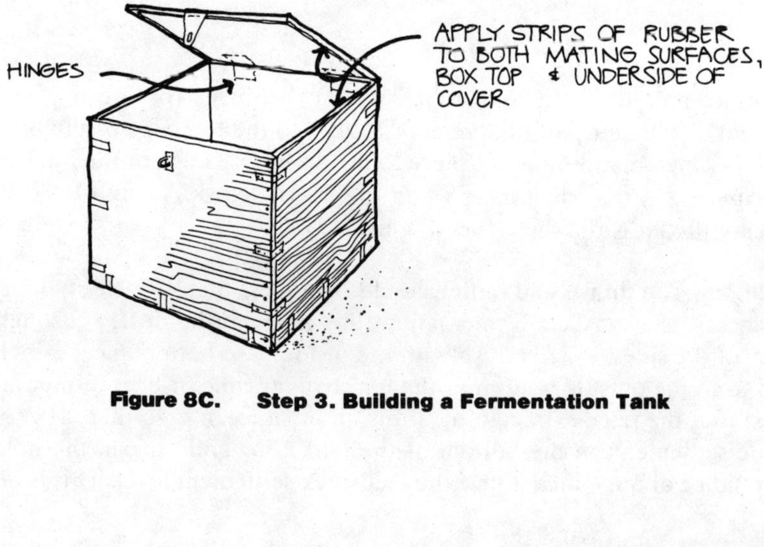

Figure 8C. Step 3. Building a Fermentation Tank

can be determined by simple arithmetic—one bushel will equal about one cubic foot—or by dumping grain into the tank to establish the approximate usage height. Multiply the dry-grain level by 1½, since the grain will swell about that much when wet. An outdoor faucet head with shut-off valve and threads to receive a garden hose can be used for this outlet. For grain mash, the bottom hole must be large enough to receive a gate valve—generally 2 to 3 inches, depending on the valve you purchase. (See Figure 8D.)

Fermentation Lock. Fermentation creates large quantities of carbon dioxide gas (two molecules of gas per two molecules of alcohol produced) that must have somewhere to go. The trick is allowing the gas to escape while preventing air from entering the tank, so you cannot simply drill a vent hole in the tank. A fermentation lock solves the

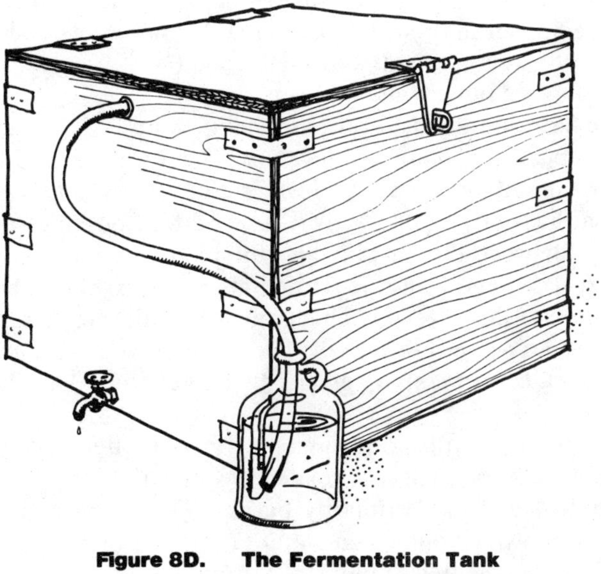

Figure 8D. The Fermentation Tank

problem by passing the gas through water, which subsequently blocks air from entering. You can buy a commercial lock that installs in the lid, but it is far easier and cheaper to build your own.

Use a 6-foot piece of plastic or rubber hose of any diameter, say up to ¾-inch. Drill a hole in any side, 1 inch down from the top edge of the tank. The hole must be slightly smaller than the outside diameter of the hose. Press-fit the hose into the hole so that it extends at least 1 inch into the body of the tank. The other end of the hose will dangle into a jar or bucket of water. As gas is formed, it will bubble its way out. During any lulls in fermentation, air will be blocked from entry by the water in the jar. The fermentation lock completes construction of your tank, unless you wish to install a heat exchanger used to dissipate heat generated by fermentation. If fermentation heat exceeds 95°F, it will kill the yeast, concluding fermentation. (See Figure 8D for details.)

Heat Exchanger. A heat exchanger is a ¾-inch-diameter copper coil spiraling down into the tank, like a spiral staircase. The bottom end of the coil is tapered and attaches to a ½-inch-diameter copper tube that goes straight over and out one of the tank's side walls and up into a reservoir of water. The top end of the coil (¾-inch) goes straight up and out into the same container.

As the fermentation heat raises the temperature of the water within the coil, a push/pull situation develops. The warm water rises up into the reservoir and pulls some cold water down; some cold water (being somewhat heavier and less active) drops down, pushing some warm water up. Between the pushing and the pulling, the water within the coil is circulated and cools the fermentation tank to about the ambient temperature of the reservoir. This is an automatic, no-energy-cost system, so it does not add to the cost per gallon of your alcohol. (Details for making a copper coil are given in Chapter Nine: Distillation.)

Options. A trip to your local junk yard may produce a low-cost fermentation tank, or at least inexpensive materials to fabricate one. Avoid hot-water tanks, primarily because of loading and unloading problems. Inlets and outlets on water tanks are generally too small to make them feasible. Large industrial drums are viable, provided they can be sealed and do not present a loading problem. An old refrigerator laid on its back makes an excellent tank; particularly in the winter when the insulative qualities will prevent mash freezing. Interior components, such as a circulating fan and light, obviously must be sealed; and of course discharge plumbing and a fermentation lock must be installed. A group of small containers, such as five-gallon plastic drums, can be connected in series by running hose from one drum to the next for gas purging. This offers high volume (ten of them gives you a fifty-gallon capacity) at low cost. Imagination pays off in time and money saved.

THE FERMENTATION PROCESS

Once cooked, the mash and broth must be cooled to room temperature. It could be immediately placed in the fermentation tank, but unless your tank is made of metal, the mash will cool faster in the cooking vessel. To expedite cooling, you can add cold water to bring the total volume to thirty-seven gallons per bushel of grain. Do not add any further ingredients until the temperature drops to below 80°F.

When the temperature is below 90°F, transfer the mash/broth to the fermentation tank, and add the balance of your malt left over from Chapter Seven (four pounds). Stir this thoroughly and let it sit while you prepare the yeast for inoculation. Keep the lid of the fermentation tank closed while preparing the yeast to avoid mold/bacteria contamination.

Yeast Inoculation. There are three basic options for yeast inoculation:

1. Add four ounces of yeast per bushel of grain directly to the mash slurry.
2. Culture yeast on a nutrient-loaded medium and add to the mash slurry.
3. Culture yeast on a small quantity of fortified mash, and then add to the main mash batch.

Yeast added directly to the mash is by far the quickest and easiest way to inoculate the mash. If you use dry yeast, you can either premix it with water—65°F to 80°F—or stir it directly into the slurry. Dry yeast should not be prepared more than one day before use, unless mixture is refrigerated. Wet yeast (chunky style) should be thoroughly mixed with enough water to form a thin soup. This soup is then added to your mash. If you crumble chunk yeast directly into the tank, it will not immediately mix thoroughly with the mash and malt, and will initially develop pockets of yeast (colonies), diminishing the productivity of your fermentation. The yeast cells must be thoroughly dispersed throughout your mash slurry for maximum efficiency.

Culturing yeast prior to mash inoculation accomplishes several things. It activates the yeast so that when it is placed into the mash it immediately goes to work. It increases the yeast population, so that you can begin fermentation with a greater number of yeast cells compared to your initial input. It gives the yeast a chance to get a few vegetative cycles out of the way. When yeast vegetates or respirates, it is not producing alcohol. Probably most important, culturing preconditions the yeast for a given mash medium. It affects the types and strength of enzymes produced by the yeast, which have a direct bearing on its ability to efficiently ferment mash. If, for example, you culture yeast in a glucose-strong medium, the yeast will become lazy and will not manufacture sufficient quantities of amylase and zymase to effectively break down carbohydrates during fermentation. In essence, you can, to a small degree, control the genetics of your yeast cells by controlling the culture medium. This is not really genetic manipulation, or mutation, but a matter of conditioning the cells in a certain direction.

A nutrition-loaded culture medium takes one day to make. If you are using a grain feedstock, use a potato starch medium, and vice versa; for a potato feedstock, use a grain medium.

For grain feedstock: Shred one whole potato, including the skin, into a 2-quart container. Add 1 quart of pure water and refrigerate overnight. The next day, bring the contents to a boil for 1 minute; cool the liquid; and filter out the solid potato particles through a clean white cloth or filter paper. To this liquid add ¼ teaspoon plant food, one multivitamin pill dissolved in 4 ounces of water, and, if available, ¼ teaspoon agar, a product made from seaweed and obtainable from a chemical supply house or from some health food stores. Do not add any sugar whatsoever. To this liquid, add 4 ounces of yeast for each bushel of grain mash you are fermenting. Cover the container with a clean cloth, and let it sit overnight. The following day, inoculate your mash.

For potato feedstock: Use 4 ounces, more or less, of canned creamed corn, and 1 quart of water. Add ¼ teaspoon plant food, 1 multivitamin pill, ¼ teaspoon agar, and 4 ounces of yeast per bushel of potatoes, and let sit, covered with a clean cloth, overnight at room temperature. The following day, inoculate your mash.

Culturing yeast in a fortified mash medium requires that you hold back or save part of a previous mash run. If you decide to use this medium, save about one quart of your mash prior to fermentation, and refrigerate until one day before you are ready to cook another batch of mash. At that time, add ¼ teaspoon plant food, one multivitamin pill, and ¼ teaspoon agar, and let sit overnight. The following day, inoculate your mash.

With the exception of Option One, you must begin yeast preparation one to two days before you actually need it. In lieu of culturing, you can take the less efficient but easier way of directly adding the nutrients at the time you charge the tank with the mash slurry, malt, and yeast.

Once the mash slurry has been inoculated, you can simply close the lid, place the gas hose into the water tank, and walk away from it. The yeast and enzymes take over and do the rest from here. Dependent upon the ambient temperature, the mash should "kick" (begin to ferment and emit gas) within two to six hours. It is best to plan inoculation at least six hours before sundown. Once the yeast takes off and the mash kicks, it will provide its own heat during that first cool night. If it has not kicked by nightfall, a cooler temperature may delay active fermentation by subduing the yeast. This does not apply if you are fermenting in the house or within a heated or warm shed or garage.

If you do not have a steady flow of bubbles from the fermentation lock at the end of twenty-four hours, and the ambient temperature is above 60°F, something is wrong, and it is time to backtrack. Open the fermentation tank and see if there are any bubbles at all on the surface. If there are none, give the side of the tank a good kick. If no bubbles are to be seen, it can be presumed that for some reason your yeast died during the night–or prior to inoculation. Forget the priest and burial ceremonies. Check and adjust the pH to 3.5 to 4.5 if necessary. Add original quantity of yeast used, and recap the tank.

If there is scum floating on top of the slurry, chances are good that it is mold contamination. Generally white or pink, it will resemble cotton candy. Skim the mold from the surface, remove several gallons of the fluid slurry, bring it to a boil, and return it to the tank while at boiling temperature. This will kill almost any mold present, but will also kill the yeast. When you boil out the mold, add original quantity of yeast used when the temperature is safe to do so, stir thoroughly, and recap the tank.

If your mash fails to begin fermentation, there is a slight possibility that you are using stale (dead) yeast for inoculation. Rather than wait another full day to see if the batch kicks, prepare a glucose solution–one tablespoon sugar in a quart of water–at 70°F to 75°F, and add one-half ounce of the yeast you used for this batch. If within one to three hours there are no bubbles, odds are that you have bad yeast. Buy fresh yeast and inoculate your batch. Save the stale yeast for your next cooking session, for there will still be a modicum of enzymes within the cells.

If, after all this and another twenty-four hours, your batch has still not kicked, it may be contaminated from your cooking vessel or utensils, from the coating used on the inside tank walls, or from your general work area–dirt, dust, bacteria, or mold. As a final check, draw a one-quart sample, bring it to a boil, reduce to room temperature, add fresh yeast, and watch it for one to three hours. If you still get nothing, the problem is most likely in your initial cooking step–you will have to lengthen your cooking time. If the problem is the coating on the tank, line the tank with plastic or recoat with another product.

When your batch finally kicks, you can check the fermentation lock now and then, but for the most part you can leave the tank alone for three days. At the end of three days, draw a sample from the upper faucet, let it sit long enough for any solids to settle out, and check the alcohol content with a low-proof hydrometer. If all has gone well, you will have anywhere from 4 to 8 percent alcohol, but do not be dismayed if you don't. Check the proof each day thereafter until you obtain a reading in excess of 10 percent. At that point, you can let it continue to ferment and shoot for a higher proof, or shut down and distill what you have. Your decision will mostly be based on time. Once the proof reaches 10 percent, the fermentation process begins to slow down. Each 1 percent proof beyond 10 percent takes about twice as long to produce. If it took five days to produce a 10-percent proof, it will take another two days to get the proof up to 11 percent; another two to make 12 percent; and perhaps three or four more days to hit 13 percent. While you are trying to increase your yield, you are also increasing the odds that acetobacter will contaminate your mash and begin to form acetic acid. If time is no problem, and you want to go for the higher yield, there is a simple early-warning test to let you know when acetic acid is beginning to form.

Each day past the 10 percent reading, draw a small sample—a tablespoonful will do—and add it to a small glass of fresh, warm milk. If the milk curdles or becomes spongy, acetic acid is present, and it is time to terminate fermentation. If there is no change in the consistency of the milk, you can continue fermenting.

At any time during the fermentation cycle, you may notice that the bubbles stop, and the process seems to come to a dead stop. If you have had active fermentation, the yeast is probably going through a vegetative cycle, and there is no real cause for alarm. When the bubbles stop, draw a sample and test the proof. If it is below 10 percent, give the tank a day to kick again. If it does not, you can conclude that the fermentation has gone as far as it will ever go, and you can opt to either conclude that batch, or add one-half cup of plant food to attempt to induce further fermentation. If you choose the plant food option, give it at least six hours to kick. If it does not, you have reached the termination point.

What to Do If You Get a Low-Proof Yield. There are a variety of reasons for a low-proof yield on any given run. Air contamination will create almost immediate and automatic yeast respiration. During respiration, the yeast will continue to feed on the glucose, and will continue to produce carbon dioxide, but instead of generating alcohol, it generates water. So, although your fermentation lock may be bubbling away, you are only producing water. On the first low-proof batch you experience, check the tank to make sure the seal is tight and that there are no air leaks.

A low-proof yield may also be caused by inadequate hydrolysis. To insure adequate hydrolysis, increase the cooking time of your mash on the next batch. (Distill the low-proof batch, which will give you the alcohol it contains and will further hydrolyze the mash. Ferment this batch one more time to get as much alcohol as possible out of your feedstock.)

On a large fermentation tank, natural and normal heat dissipation may be lacking. The heat generated by fermentation within the first day or two may be killing the yeast. Check the ambient temperature on the next batch, and if it exceeds 94°F, you will either need to go to a smaller tank or install a heat exchanger on the tank. (See How to Build a Fermentation Tank, page 82.)

When all problems have been solved, and all else goes well, and you have the highest proof you are going to get, it is time to discharge the tank. With a grain feedstock, this is a two-step operation: discharging the upper fluid, which is extremely easy, and discharging the lower mash, which takes a little time and effort.

If the tank sits higher than your still, you can run the hose right into the still and open the faucet. If not, you can drain the upper fluid into buckets and manually charge the still.

On the first mash run, there are several options for handling the mash: You may leave it in the tank for a second run; you may drain and filter the mash, returning it to the tank for a second run; or you may drain and filter the mash and sell it to a farmer.

The first option is expedient and easy. Any alcohol in the mash will not be lost; it will be diluted by the addition of fluid for the second run, and will be picked up in the final distillation run. The second option is time-consuming, but you acquire more beer for initial distillation. (Beer is the term used to describe the mash once it has fermented.) However, you must again drain and filter at the conclusion of the second run–double handling. The first and second options involve time factors. It will take about three days longer for the second run than for the first, with a resultant lower yield. If it took only three to five days for the first run, it might then be worth it. If, however, it took seven or eight days for the first run, it may not be worth tying up your equipment to squeeze out that last drop of alcohol from your feedstock. The third option is simply deciding to conclude that particular run, and start a fresh batch.

Filtering can be accomplished by placing hardware cloth (¼-inch mesh) over a container, placing plastic window screen over that, and emptying the mash onto the screen. The filtrant (beer) is then distilled. This can be modified by building a V-shaped chute that leads right to the still. With cloth and screen on top of the chute, the mash is fed onto the screen; the beer runs down the chute into the still; and the mash is salvaged from the screen. The mash can then be sun dried, oven dried, or sold wet.

You do have the option of using both mash and beer for distillation at the same time. The major disadvantages are handling (getting the mash into and out of the still), inevitable scorching of some of the mash (wasted product), and, if you choose to make a second fermentation run, getting the mash back into the fermentation tank (more work and more time). A minor problem here, too, is baffles. Baffles make emptying and cleaning the still after use more difficult when chunky mash is involved.

The advantages of mash/beer distillation are that it kills any mold or bacteria present, further hydrolyzes the mash for a second fermentation run, and gives you more alcohol for that one run—the alcohol that would have remained with the mash if you left it in the fermentation tank.

The Second Run. The fluid removed for distillation must be replaced. This can be done either by adding fresh water, or by distilling the beer, cooling the distillant, and using that. The advantage to using the distillant is that even though it has been heated to near boiling, it will contain small amounts of sugar, protein, and trace elements beneficial to the second run.

Since fermentation creates certain acids, check the pH and if it has dropped below pH 4, adjust it with alkali to pH 4 or pH 5.

Add 1 teaspoon of plant food per bushel of grain to recharge the phosphate in the mash and help kick the second run. Add 2 or 3 multivitamin pills dissolved in a glass of water (about 8 ounces).

Regardless of precautions, it is inevitable that some acetic acid will be formed during the first run. Use the warm-milk test, and if the milk curdles, add calcium carbonate to the mash to buffer the effects of the acetic acid. The quantity of carbonate depends upon the severity of the acetic acid. Add a small amount and retest for the acid. When satisfied, cap the tank.

If the tank does not kick within six hours, add 2 to 3 pounds granulated sugar to literally wake up the yeast, and recap the tank. For the next several days, test the alcohol proof and test for acetic acid. If the proof does not increase and the acetic acid does, scrap the run and salvage the mash for sale to a farmer. At this point it must be presumed that bacteria has a good toehold on the mash, and if so, it is all downhill from here. Any nitrates present may be reduced to ammonia; any sulfates present may be oxidized to hydrogen sulfide, which is toxic to respiratory organisms.

SEASONAL CONSIDERATIONS

Fermentation is hardest to accomplish during the winter months, since yeast becomes inactive at temperatures below 60°F. If your fermentation tank is outside or in a nonheated building, it must be well insulated. The ambient temperature of the fermentation tank should be 80°F to 85°F when inoculated, to give the yeast a chance to take off. Once fermentation begins, it will generate its own heat and survive, provided that heat is not too rapidly dissipated. Even so, total fermentation time is increased during the winter months.

The fermentation tank can be heated with warm coils, provided the temperature does not exceed 94°F. This heat can be obtained from your regular home heating system, if you have hot water heating.

The distillation fuel cost per gallon of ethanol produced is higher in the winter, for it takes more heat to raise the beer temperature from a charge level of, say, 30°F than from summer temperatures of 70°F, to 90°F. In this instance, a cogeneration system is advantageous, for it uses heat already generated for the primary boiler to fire a secondary unit. (More on this in Chapter Nine: Distillation.)

Solar still efficiency is greatly diminished during winter. A solar still will work provided there is sunshine, but it takes a lot longer and the proof is much lower when the temperature drops.

The major advantage to wintertime production is that you can increase the base proof yield of your runs prior to distillation, and eliminate acetic acid from your fermentation tank(s). Acetic acid freezes at 62°F, water freezes at 32°F, and alcohol freezes at −173°F. This is especially great for multiple fermentation runs, for you can literally purge one run of acetic acid prior to starting the next run.

By exposing the completed fermentation run to Mother Nature's freezing touch for a day or two, you will cause almost all acetic acid to crystallize, along with much of the water. However, the alcohol, acting somewhat as an antifreeze, will inhibit total freezing of the undesirable water and acetic acid.

Once crystallization occurs, the ice and slush can be removed mechanically—by hand or sieve strainer. The liquid left behind can be set aside for distillation. If the weather is below freezing and crystallization does not occur within several days, kick the tank or container, and crystallization should occur almost immediately.

A minor advantage to winter is that activity diminishes—social, farming, and recreation. This releases more time for alcohol production. Nothing better to do? Make some alcohol.

Distillation condensation efficiency is increased during winter. Once the boiler reaches operating temperature, the balance of the system will be cooler, resulting in more favorable alcohol condensation.

9
Distillation

HOW IT WORKS

IN SOLUTION, regardless of the percent alcohol by volume, the carbon bonds of water and alcohol are very strong. It is literally impossible to separate them by mechanical means (centrifuge, osmosis, etc.) even though their respective specific gravities are different. It *is* possible to add chemicals to the solution to assist in distillation, but they serve as a catalyst at best, are sometimes hazardous, and do raise the total cost of the alcohol produced. It is also possible to utilize a filtration system during distillation to increase the final proof alcohol yield (more on this later).

The basic principle that makes distillation work is boiling point—water boils at 212°F and alcohol at 173°F. Were it not for a simple law of physics, the temperature of a heat-fired still would run up to 212°F and boil off both the water and alcohol, so that the resultant condensed vapors would still be water and alcohol. However, when heat is applied to the still, the temperature rises to the lowest boiling point (alcohol, 173°F) and is maintained as long as alcohol is in solution. The reason the temperature does not increase to the boiling point of water is kinetic energy reduction. As the faster-moving molecules near the surface of the solution vaporize, they lower the kinetic (active) energy of the remaining solution. When the alcohol has vaporized out of solution, the temperature begins to rise to the next highest boiling point until it reaches 212°F and water begins active and rapid vaporization.

This is not a completely segregated process. While alcohol is vaporizing at 173°F, some water molecules escape with it (about 20 to 25 percent of the vapors will be water), even though the boiling point of water is 212°F. The trick is to minimize the amount of water that actually makes it through the condensing phase of distillation, and that subsequently lowers the final proof of liquid alcohol. To a certain degree, this can be accomplished by efficient still design, and this can mean the difference between 150 to 160 proof alcohol and 190 proof alcohol, the latter being the more favorable for any planned use of the product.

THE BASIC STILL: WHAT IT'S MADE OF AND HOW IT WORKS

Anyone who can boil water can distill alcohol. The most difficult aspect is setting up (building) the equipment once the basic still design is chosen. The most important consideration is efficiency, which is controlled by fuel source and cost, and equipment design and process factors. Ideally, you will maximize both factors to keep your per-gallon cost to an absolute minimum and your proof yield to a maximum.

The simplest heat-fired still is a "moonshiner" still (see Figure 9A) consisting of:

1. Boiler: This holds the liquid for heating to vaporization point. Made preferably of metal, it can be almost any shape–round, square, rectangular.
2. Cap. This serves as the initial condensing unit for water vapor caught up with alcohol vapor.
3. Intermediate condensing unit. This serves to precondense any water vapor that makes it past the cap.
4. Condensing coil. This is used for the final-stage condensation of alcohol vapor to a liquid form.
5. Flakestand. This cools the condensing coil as hot vapor passes through it.

Boiler. Copper and stainless steel are the preferred metals for still construction because alcohol is less reactive with these two metals than with other metals, thus yielding a purer end product. Of the two, copper is preferred for its malleability and cost. Stainless steel is more difficult to work with and more expensive. Carbon steel is the next most favorable choice. Although less malleable than copper, carbon

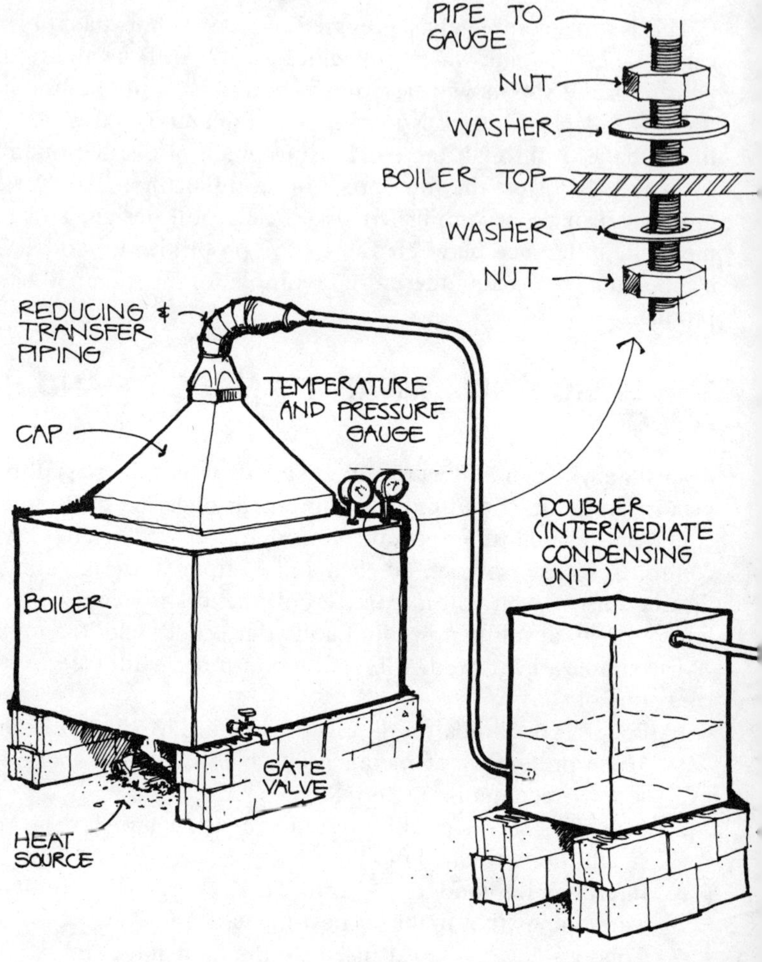

Figure 9A. Basic Still Design

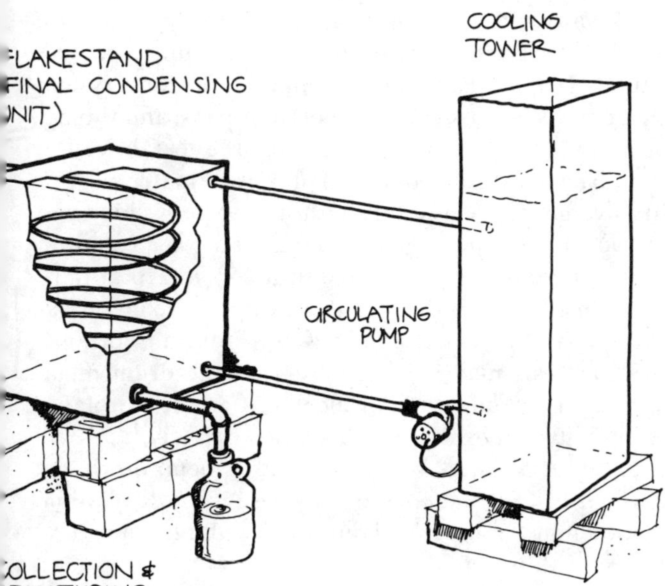
COOLING
TOWER
FLAKESTAND
FINAL CONDENSING
NIT)
CIRCULATING
PUMP
OLLECTION &
ENATURING
ONTAINER

steel is by far less expensive. The major disadvantages of carbon steel are rust and corrosion, which can be somewhat offset with proper care and attention. Cast iron is worthy of consideration although it is difficult for the average person to work with. Many industrial and commercial heating systems utilize cast-iron boilers. Malleable cast iron is easier to work with, but considerably more expensive. Aluminum deserves the least consideration. Welding aluminum is an art—extremely difficult for the novice. Aluminum is also highly reactive with alcohol, and although it will not rust, it will pit and corrode. Galvanized metals may be considered, but caution is advised during fabrication of the boiler. Welding galvanized metal produces toxic fumes, so welding should be done in a well-ventilated area.

Where cost is a factor, current market conditions and prices will have a major influence on your choice of boiler material. Before you make a final decision, shop around.

Junk. For the junk yard enthusiast there are numerous possibilities. Consider discarded water heaters (glass lined and set up to receive electricity or natural gas), workable and reworkable heating system units, and assorted barrels, drums, buckets, pots and pans, and tubing. With a little ingenuity and imagination it is possible to fabricate a reasonably efficient back-yard still from local junk-yard materials. For example, two fifty-five-gallon steel drums (industrial type) welded end to end will allow you to produce about seven gallons of alcohol on one distillation run. These drums generally sell for $5 to $10 at most junk yards. The intrinsic disadvantage of these drums is contamination. You have little way of knowing what was in the drum prior to its being discarded. Chemicals may have ionized with the drum metal and may subsequently contaminate your alcohol. Certain petroleum products will be literally impossible to remove completely, and will contaminate your product. Unless you can see bare metal inside the drum, do not consider it. Even then, before you use it for distillation, boil a caustic solution in it for about an hour, rinse with bleach several times, and then rinse with water.

Fabrication. Although rivets and bolts can be used to fabricate a boiler and related components, welding and/or soldering are more preferred, both for durability and efficiency. There is also a safety factor to be considered: A seam or joint that leaks could cause an explosion. So, if you plan to build your own, resign yourself to welding or soldering.

Soldering is much easier and the cost of equipment less expensive than for welding. If you choose copper as your boiler/still metal, you can easily get by with soldering all seams and joints. A basic soldering kit–blowtorch, solder, and flux–can be obtained for under $30. One or two hours' practice on scrap copper will just about qualify you as an expert. Although steel can be soldered with modest success, welding is the more desired method of attaching parts. A small do-it-yourself welding kit can be purchased for under $100 at many chain discount stores. It takes a little longer to develop welding proficiency than it does to develop soldering ability, but it is not impossible for anyone with the desire to learn.

The Cap. The cap sits squarely on top of the boiler and while its presence is not absolutely necessary for the still to function, it does increase efficiency. The cap can be of vertical side-wall design, which is easier to fabricate, but it is far more efficient if the walls are constructed at a 30- to 45-degree angle to the boiler. Basically, vapors rise straight up until they hit something. With vertical walls, surface area contact is restricted to the top of the cap, where condensed vapors cling until they are heavy enough to fall back into solution. An angular design allows vapors to begin surface contact and condensation almost as soon as they leave solution, and rather than drop back into solution, they slide back down upon condensing. This means that more water slides down and more alcohol vapor escapes into the next condensing phase. More work to fabricate, but more efficient in terms of productivity.

The cap can be conical or pyramidical (four or three sided), but more important, it need not be of the same material as the boiler. If you choose steel for the boiler, you can make the cap copper. The major advantage of a copper cap is weight. Even if you pipeline your distillant into the boiler, you need easy access to the boiler unit for cleaning and occasional repairs. Although the actual weight involved is governed by the size of your equipment, copper is definitely lighter than steel.

Intermediate Condensing Unit. More commonly called a thumper or doubler, this unit phases out some of the water that made it past the cap. It is far enough removed from the boiler so that its heat source is the heat of the vapors (173°F) and therefore it has a greater tendency to condense water vapors and let the alcohol pass through. Even so, some alcohol will remain in the doubler after your distillation run, but this can be drawn off and put into the main boiler on your next run. In a sense, the doubler serves as a boiler with an extremely regulated temperature, and it does increase the efficiency factor of your still by about 10 to 15 percent.

The size of the doubler is important for two reasons. If it is too small, condensed water vapors will fill it and be pushed into the final condensing coils, resulting in a lower-proof (polluted with water) end-product alcohol. If it is too large, the heat will dissipate, forcing the ambient temperature to drop below 173°F and subsequently condensing alcohol back into solution. Regardless of the size of the main boiler, the doubler should be no less than 20 percent of its size and no more than 50 percent. Capacity and heat problems can be overcome by using wood to construct the doubler. Wood will not dissipate too much heat, is easy to work with and make adaptions to, will not appreciably contaminate the alcohol, and is not costly to replace as a component part of your equipment. In fact, an old wooden keg makes an ideal intermediate condensing unit.

Condensing Coil. This is the final stage of your active distillation process, and consists very simply of copper pipe wound into a coil much as a snake coils in the sun. Although this pipe could well be straight and run through a long water jacket, coiling saves space and facilitates cooling. Copper is the preferred metal, for it is soft and easy to bend without kinking and binding.

The diameter of the pipe and length before coiling are the two most important considerations. If the diameter is too small, clogging may occur, causing back-pressure into the boiler, and resulting in rupture of the vessel or other component part. Anything smaller than a half-inch inside diameter (I.D.) should be avoided. A pipe with an I.D. larger than three quarters of an inch becomes cost-prohibitive and less malleable, for the larger the diameter at this point the more brittle the metal composition, i.e., the less workable and bendable.

If the length of the pipe is too short before coiling, there is not enough surface area for adequate condensation of vapors, and alcohol is lost to the atmosphere in gaseous form. A twelve-foot length (for ¾-

inch pipe), or eighteen-foot length (for ½-inch) is adequate provided you use a flakestand to cool the coil.

Flakestand. The only function of the flakestand, which is basically a cooling tower, is to keep the coil cool enough to condense alcohol vapors passing through. Although it can be constructed of wood, metal is preferred for heat dissipation. During distillation, the ambient temperature of your entire system, including the water in the flakestand, will reach 173°F. This can be offset by circulating the water, replacing the water by the addition of fresh (cool) water, or by using a heat exchanger to remove heat from the water as the temperature rises. (More on this later.)

OTHER COMPONENTS

Although a rudimentary heat-fired still will work well, you may want to add a few pieces of equipment both for safety and for increased efficiency.

Blowout Pressure Valve. Not a hundred-percent essential, depending on the size of your coil and the probability of a line clogging, a blowout valve is strictly a safety item. If a line does become clogged, the pressure will increase to the point of blowing up your still. This valve prevents that. Even if alcohol vapors do not explode, the steam, hot liquid, and ruptured metal present enough of a hazard. This valve can be obtained from almost any heating and plumbing outlet, or from a discarded water heater.

Pressure Gauge. An early-warning device, it works for you only if you keep a steady eye on it. As long as a line does not clog, the pressure within your system will be literally zero. If you use a pressure gauge, and the pressure begins to rise, shut down immediately, even if you also use a blowout valve. The cause of the problem will, in most instances, be a clogged line, which will have to be cleaned out. If you let the pressure increase to the point of activating the valve, you will waste either alcohol vapors or liquid. Once the problem has been solved, distillation can be continued.

Temperature Gauge. Another early-warning item, the temperature gauge, announces the conclusion of your distillation run. When the temperature lifts off of 173°F, almost all of the alcohol has left solution and water will begin to dominate the vapors. Without a gauge, you can tell when the run is finished by watching the end of the coil. When the stream of liquid alcohol stops flowing—an interruption that lasts only as long as it takes the boiler temperature to rise from 173°F to 212°F—the run is complete. Thereafter, continued operation will be distilling only water, and if collected with the alcohol, this water will lower the proof of your final product.

The installation of temperature and pressure gauges must be made so that they do not come into direct contact with liquid. They can be installed directly on top of the still, or on a side, close to the top. You are not concerned with the liquid temperature, but rather the vapor temperature. Likewise, if something goes wrong, you are concerned with the vapor pressure, not the liquid pressure. By the time a pressure gauge indicated a liquid-pressure change it would probably be too late—something else would have erupted first.

Although not absolutely necessary, the blowout valve should also be placed so that it has no direct contact with liquid. If it is in contact with liquid and it blows, there will be a geyser effect and hot liquid will spew from that side of the still.

HOW TO BUILD A BASIC STILL

The Boiler. Regardless of the metal chosen for fabrication, the bottom plate of the boiler should be twice the thickness of the walls and top, primarily for durability and longevity. The bottom plate is extremely subject to burn out because it receives the direct blast of fire from your heat source. An alternative is to construct the entire boiler with the same gauge metal, then set it on a sheet of plate steel. The plate steel serves to buffer the impact of direct fire, and transfers heat received to the boiler.

For a decent-size boiler (say 750-gallon capacity) of square-base design, begin with a 5′ × 5′, ⅛-inch base of sheet steel (¼-inch if you do not plan to use a buffer plate), placed on its final firebox resting place—you will not want to move the boiler once construction is complete simply because of the weight. Build two 5′ × 5′ wooden collars (construction-grade two-by-fours will do). These collars—one for the top, one for the bottom—will hold the sides in place both squarely and vertically during welding. A third collar may become necessary if your sheet metal buckles, forming a gap on any corner edge (see Figure 9B). Placed halfway up the boiler, it will pull the edges into flush alignment. This phase of construction usually requires two people.

The corresponding size for the walls is: two pieces 4′ × 5′; and two pieces 4′ 11¾″ × 5′. The front and rear walls are then placed on top of the base first, and the smaller side walls inserted between them. The quarter-inch variance makes up for the thicknesses of the front and rear walls—⅛ inch each. The perimeter of the tank is now 20′.

Once pieces are in position, with collars in place, spot-weld the bottom, middle, and top on each of the four walls. If a problem is to develop now, you will not have to rip out a full seam weld, but can break the spot welds for corrections. If everything looks good thus far, weld all corner seams full length from the bottom up. (Since heat rises, going from bottom up will preheat metal about to be welded.) All welds are done on the inside joints.

When all four corners have been welded, remove the collars and check wall squareness against the base plate. If there are no gaps, weld the inside perimeter full length around where the walls meet the base plate. Let all welded seams cool as slowly as possible—do not

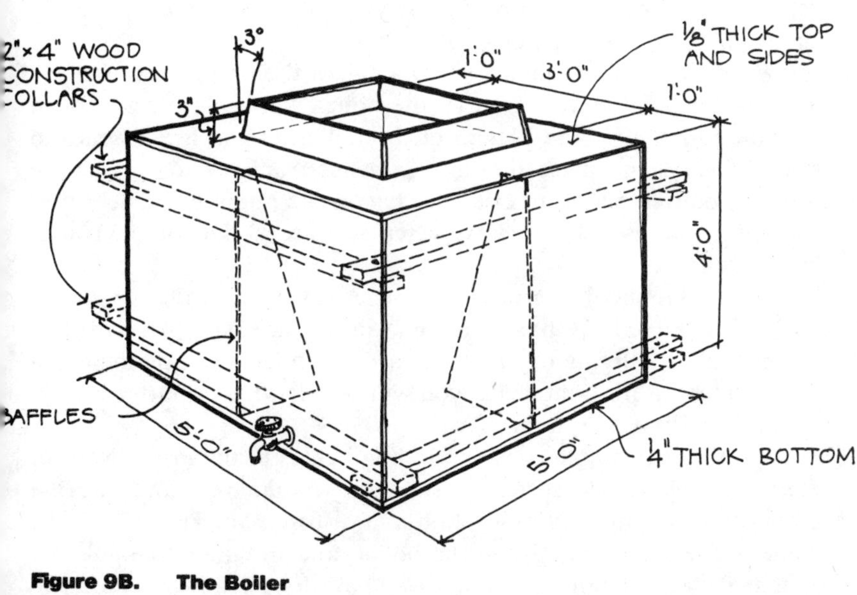

Figure 9B. The Boiler

quench with water or oil. The longer it takes for the seams to cool, the stronger the joints will be.

When these welds have cooled to the touch, fill the box with water to check for leaks. If there are leaks anywhere, you will have to reweld that spot as well as the seam line on either side of it for at least 3 inches.

If there are no leaks, it is time to cut three holes in the walls: (a) one to charge (fill) the boiler, (b) one to purge (empty) the boiler after distillation, and (c) one for the blowout valve installation. A hand-drill attachment that receives hacksaw blades of various diameters (available at most hardware stores) can be used to cut these holes. Here you have two options: Drill before filling the box with water and lubricate the saw blade with oil; or drill after filling, letting the water in the box cool the metal as you drill. The disadvantage to the former is that you must cap the holes after drilling in order to fill the tank for the leak test. The disadvantage to the latter is that while you are drilling, water will begin leaking out and onto the drill, perhaps causing a short circuit. For the sake of safety, the former is more advantageous.

The recommended size for the charge hole and the purge hole is a minimum of 2 inches. The size of the hole for the blowout valve depends upon the diameter of the blowout-valve feed pipe, but generally, a hole 1 inch in diameter will suffice.

If your fermentation tank sits higher than the boiler, and you use the principle of gravity to charge the boiler, that hole can either go near the top or near the bottom of the boiler. If you use a pump to charge the system, place the hole 3 inches from the top. Charging from the bottom places an extra burden on the pump by having it fill the boiler and also work against water pressure as the volume (water level) rises.

Upper holes should be placed 3 inches down from the top of the boiler. The purge hole (b) should be flush with the base plate to allow complete fluid purging of the boiler after distillation. If the hole is a quarter inch up from the plate, you will be left with a quarter inch of liquid that will not drain out.

You may at this time consider adding baffles to the inside walls of the boiler. These add structural strength to your unit, and increase efficiency by aiding heat absorption and distribution. They should be made of the same material as the boiler, and in either triangular or rectangular shape, with a 1-foot base and 4-foot height. Welded two each to the front and back walls, and one each to the side walls, they will absorb, by conduction, heat from the base plate and walls and transfer that heat energy to the liquid.

Although a rectangular baffle will use twice as much metal as a triangular baffle, you will have that much more surface to capture and distribute heat. The rectangular baffle will also serve as a base upon which the top can rest.

With or without baffles, you need a top on the boiler to receive the cap. The top consists of two sheets 1′ × 5′, and two that are 1′ × 3′, ⅛-inch thick, laid to form a perimeter shelf (top) 5′ × 5′ outside dimension, and an access hole 3′ × 3′. Once the top is welded in place a 3-inch-high flange, 3′ × 3′ with a 3-degree inward-pitched taper, is welded to "window-frame" the access hole. The taper is designed to snugly receive the cap. The corner edges of the flange should be rounded with a file to facilitate cap fit.

The Cap. This can be made of sheet copper or steel, 1/32 inch (.0312 inch) or up to 1/16 inch (.0625 inch) thick. The exact dimensions will vary slightly depending upon the angle (pitch) you place on the cap—from 30 to 45 degrees—and the desired top-of-cap opening. Basically

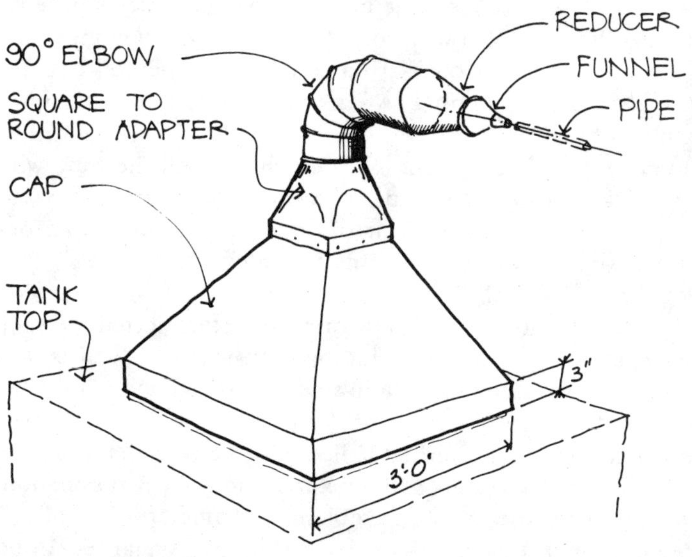

Figure 9C. The Cap

you are constructing four equilateral triangles, each with a 3″ × 3′ flange on the bottom, and a 2-inch flange on top (see detail, Figure 9C). A heating-flue adaptor–square on one end and round on the other–and a 90-degree elbow can be used to top off the cap. A flue reducer–large on one end tapered to small on the other end–can be used to accommodate the smaller plumbing leading into the doubler–1- to 2-inch piping.

The Doubler. The important consideration on the doubler is not its composition–wood or metal–but that the charging pipe carrying the alcohol/water vapors enters the doubler from the bottom, and that the discharge pipe carrying predominately alcohol vapors exits from the top. If the vapors charge from the top, efficiency is diminished, for the vapors will take the shortest path and exit loaded with water. It is imperative that the vapors pass through liquid in the doubler heated to 173°F for intermediate condensation to effectively remove water vapor.

If you use a wooden doubler, it should be painted inside and out with two coats of an alcohol-base primer to protect the wood, prevent swelling, and avoid alcohol absorption by the wood fibers.

The Condensing Coil. A little care and patience are required when fabricating the coil. Even with soft copper tubing, if you move too fast at the wrong moment you will have a kink in the coil that will be literally impossible to remove. A kink in the coil represents a potential clogging problem.

The easiest method of winding the coil is to fill the tube with dry sand and slowly wrap it around a tree stump or other solid cylindrical object. The round pillar on your front porch will not work, for once the coil is wound, you must be able to remove it.

If you notice a kink developing while winding, stop immediately and heat that section with a blowtorch to further soften the metal. If that does not work, continue winding, and when the coil is finished, cut 2 inches of that section out and solder a connection joining the two ends.

The diameter of the finished coil should be considered before you begin winding. Too small a diameter inhibits heat dissipation; too large a diameter poses a space problem. A diameter of 1 foot to 2 feet is adequate. Each 1 foot of diameter consumes 3 linear feet of tubing.

The Flakestand. A nonpressure system, this can be a simple tub of water, open at the top, with a hole in the bottom for fresh water to come in and a hole at the top for warm water to leave. Naturally, you need a hole at the bottom for the coil tail to exit.

During distillation, every component part of your system will eventually reach 173°F, including the water in the flakestand. It is therefore necessary to recirculate flakestand water or add fresh water to maintain a temperature conducive to alcohol condensation. This can be accomplished manually with or without gravity feed, or automatically using an aquastat (heat sensor/regulator) to actuate a pump. Set the aquastat on 130°F start and 110°F stop. When the flakestand ambient temperature reaches 130°F, the aquastat will actuate the pump and feed fresh water into the container until the temperature drops to 110°F, at which time the aquastat will shut the pump off.

The fresh-water inlet absolutely must be on the bottom of the tank. If it is at the top, it will frustrate the aquastat with cool water, and create a thermal layer—warm water on bottom, cool water on top—which will hinder efficient alcohol condensation.

The reserve-water holding tank should be about three times the size

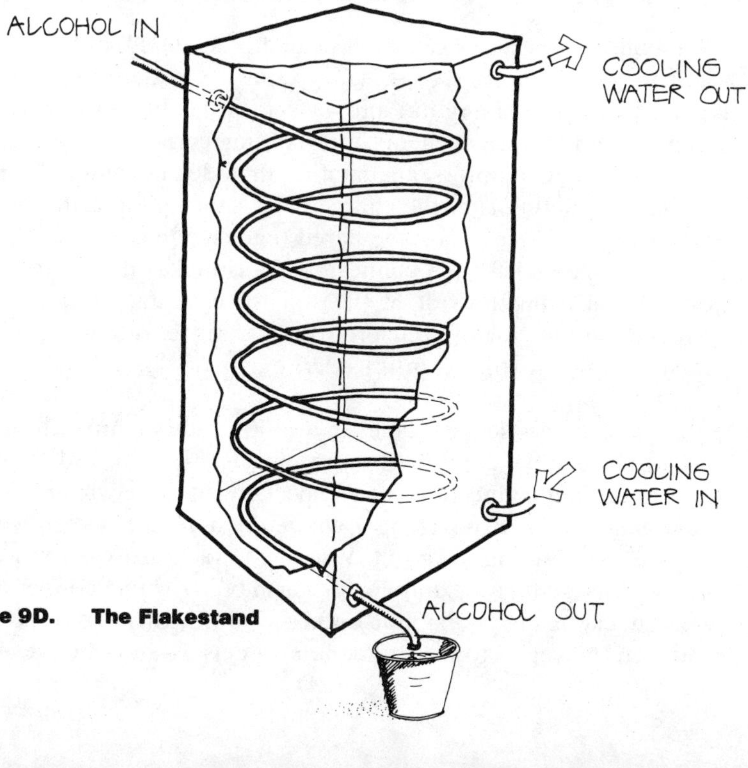

Figure 9D. The Flakestand

of the flakestand. A 1½-foot-square by 4-foot-high flakestand will hold 18 linear feet of coil wound 1 foot in diameter–six coils spaced 6 inches apart. The flakestand capacity is 9 cubic feet, or 67 gallons (9 cu.ft. × 7.48 gals. cu.ft.). Multiplied by 3, this means a reserve-tank capacity of about 200 gallons.

To facilitate cooling flakestand water displaced by recirculation, the water can be passed over a V-shaped aluminum trough prior to reentry to the reserve tank. Heat dissipation will occur in two ways: by exposure to the surface metal and by exposure to the atmosphere (some of the water will evaporate, cooling the remaining liquid form). The water will also pick up the environmental ambient temperature.

The coils should be placed in the flakestand spiraling downward like a spiral staircase. If they are placed horizontally, some alcohol will remain in the extreme bottom of each coil when the distillation run is complete. The flow progression of the alcohol vapors should also follow a warm-to-cold path–top of flakestand to bottom–for efficient condensation.

PLUMBING AND OTHER CONNECTIONS

Plumbing connections can be threaded or soldered, although pipes leading into and out of fixtures (for example, the doubler) should be threaded with a retaining nut and gasket (or flexible washer) on both the inside and outside of the fixture. Turning corners is made easy by using elbows and adaptors, again either threaded or soldered. The exception to soldering is at the charge end of the coil and the line that feeds it vapor. These should be flared (ends made conical) to receive threaded male and female connectors. This makes it easy to disconnect the coil from the rest of the still for cleaning, storage against theft, and, in the case of your prolonged absence, storage against the illegal use of your still by an intruder.

Silicone. Liquid silicone (comes in a tube), with a temperature tolerance of up to 450°F, can be used in place of washers and gaskets to seal fixtures and connections. It is especially effective where a flat nut seats against a curved surface, or where the hole size is slightly larger than the pipe leading in or out. Where contraction/expansion differentials are present–for example, in a steel boiler with a copper cap–it is resilent enough to take up the difference during distillation. Caution is advised, however, for the silicone also acts as an adhesive. If you run a bead around the cap after installation, you will more than likely end up cutting the silicone to release the cap for removal.

OPTIONS

Boiler. To increase the heat flow efficiency of your boiler, install an insulated apron to keep as much heat as possible around the side walls. The apron can be made of thin sheet metal, and should taper from 2 feet away from the side walls at the bottom of the boiler to 1 inch away, stopping 6 inches from the top. If you use an apron, you should weld a 6-inch, 45-degree-angle flange around the perimeter of the boiler 3 inches down from the top. This flange will deflect heat outward and away from the cap. Too much heat on the cap will hinder initial condensation of water vapor, i.e., the cap's ambient temperature will rise above 173°F and allow excessive water vapors to continue into the doubler.

Coil. The fabrication option here is to use 1-foot lengths of copper tubing connected with 90-degree-angle elbows in spiral-staircase fashion–same total length as the circular coil, same spacing between each section (6 inches). The disadvantages are that you will have a higher cost (each elbow), and more work (each elbow represents two soldering operations), and no more efficiency than with the circular coil.

Flakestand. If you choose not to use a flakestand, you must triple the length of your coil–which triples your material cost–to provide adequate condensation potential. In lieu of a flakestand, you can use a nearby stream as the cooling unit. With the still upstream, lay out a maze of straight copper tubing with the final collection point of liquid alcohol downstream. The caution here is that the EPA and the BATF may frown on your still being so close to a stream. They may fear the discharge of boiler effluent into the stream. Before you build, get their sentiments on your plans.

Cogeneration. Somewhat elaborate in design and construction, it is possible to build two boilers, using the vapor heat from the primary boiler as the heat source for the secondary boiler. You still need a doubler and final-stage condensing coils, but the efficiency and output of your operation is tremendously increased.

Here, the changes begin after the cap of the primary boiler. You will need a coil system that runs throughout the secondary unit, charges through its bottom and discharges from its top. This provides more than adequate distribution of vapor heat throughout. Vapors at 173°F run through the coil and heat the secondary boiler to 173°F, subsequently kicking off distillation.

Once the vapors pass the secondary boiler, they continue to the doubler for intermediate condensing, and then on to the coil/flake-stand for final condensation.

With a bit of fancy plumbing, you can run both boiler lines through the same thumper and final-stage coil. However, the primary boiler will conclude distillation much sooner than the secondary boiler, so you must include a primary-boiler-bypass purging or diverting valve between the secondary unit and the thumper. This allows "secondary" alcohol vapors to enter the thumper and excludes "primary" water vapor, which would otherwise diminish your final proof alcohol if allowed to continue through the system.

It would seem immediately apparent that if you must continue to fire the primary boiler after the conclusion of its distillation run in order to generate the steam necessary to conclude the secondary boiler's run, there is little energy-balance benefit to this type system. Not quite so. It takes more energy (Btu) to bring a fluid to its boiling point (or simmer) than to keep it boiling (or simmering). Precondensation of "primary" alcohol vapors in the secondary unit trades off energy to the secondary unit that would otherwise be lost. It also begins to bring the secondary unit into "phase" for a zero net-energy cost. Once the primary concludes, and the secondary begins, it takes literally half the energy to keep things going than if each boiler were fired separately, i.e., cogeneration consumes about 70 percent of the total net energy input of firing each boiler individually.

A long-term benefit to cogeneration is boiler life. The primary boiler, in direct contact with fire, will eventually burn out (or the plate upon which it sits will need replacement). A steam-fired boiler will not need replacement of component parts because of burn-out since there is no high-heat fire on any parts.

SELF-CONTAINED SYSTEM

It is possible to do it all in one container. You can cook and hydrolyze your grain, ferment it, and distill it right in the still. This saves processing time and space, and initial equipment material cost and construction time. Bear in mind that the major drawback to doing this is scorching the grain that is in direct contact with the bottom of the still—both during hydrolysis and distillation. It also means that you can perform only one function at a time because you are using only one piece of equipment. With separate components, you can be cooking one batch (A) while another batch (B) is fermenting, and thereafter load the fermentation tank with the cooked mash (A) while the other fermented batch (B) is being distilled.

If you choose to use the still as a do-it-all-in-one system, you must make a few minor changes in your still design. You must have some way of closing off the condensing units so air does not enter the still during fermentation. This can be done by installing a simple on/off valve in one of your smaller condensing system lines. And, you must install a fermentation lock. An option here is to run the end of your condensing coil into a bucket of water and use the entire condensing system as a fermentation lock. You will also need a faucet valve on the side of the still for drawing samples during fermentation. Unless the still is located within a warm enclosure, it must be insulated during the winter months for fermentation.

THE COLUMN STILL: WHAT IT'S MADE OF AND HOW IT WORKS

Somewhat more elaborate, but more efficient, is a column still with or without a rectifier. Basic distillation principles still apply, although the condensation process is considerably different. There are two basic types of column still, packed column and plate column, both of which can be adapted to the boiler described above. Rather than a cap on top of the boiler, a square or round column sits on the boiler and serves as both the initial and intermediate condensing unit. Unlike the cap with the preferred angular sides, the column has straight sides.

The Packed Column Still. This is the easier of the two to construct, and can be packed with almost any solid nonabsorptive material. Chunks of ceramic, glass spheres, or scrap metal are most often preferred. As vapors rise through the column, the packing material (say, glass spheres) reaches an ambient temperature of 173°F throughout the column. Because of the high surface area volume, all but about 7 to 10 percent of the water vapor condenses out and dribbles back into solution. The alcohol vapor naturally makes it past the pack for final condensation, through either condensing coils or a rectifier.

A packed column can be fabricated in one piece (a plate column still is usually made in sections) so that construction time is less and cost is much lower. It can be made of any material, including wood, for the packing material serves as the condensing medium. The proof yield of a packed column will be about 180 to 190, depending on the height of the column and total surface area of the packing material.

The diameter of the column is not that critical–it can be 6 inches to 6 feet. However, the smaller the diameter, the taller the column must be. Here, there is a too short and a too tall. If the column is too short, a lot of water vapor will pass the top, diluting the alcohol vapor. If it is too tall, energy is wasted in charging the column to an ambient 173°F.

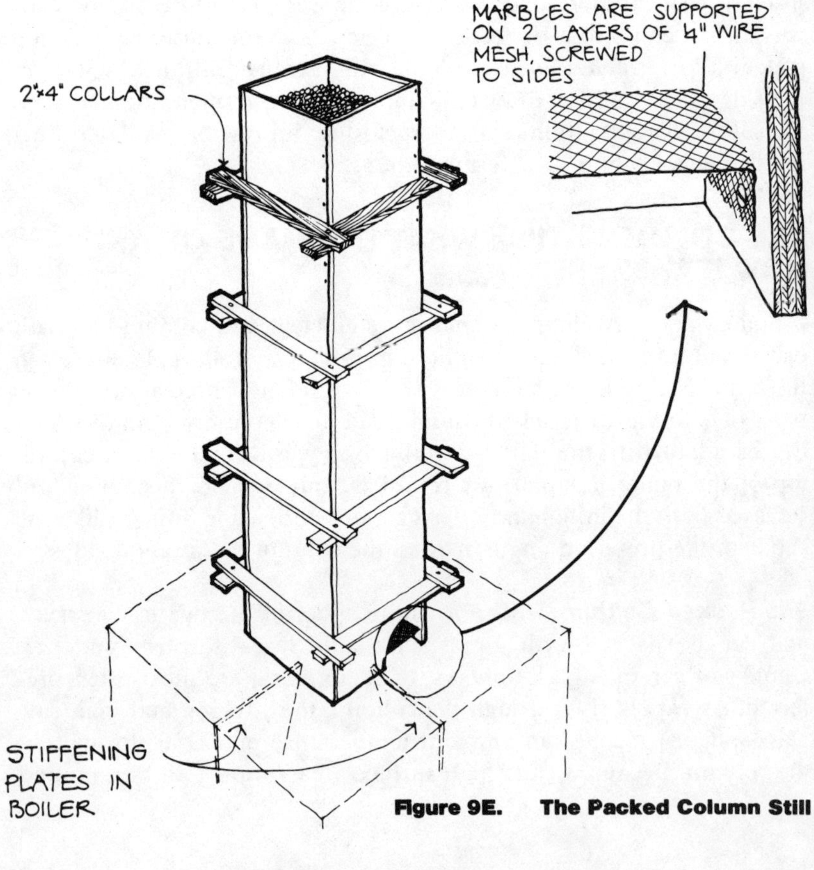

Figure 9E. The Packed Column Still

Alcohol vapor will condense back into the boiler until the entire column is charged. As a rule of thumb, the column height should be no less than 4 feet, and no more than about 10 feet, with a 1-foot diameter (or 1 foot square).

Weight is also a key factor. Once packed, the column will be very heavy. If you use a square, flat-top boiler, abutments must be included as well as angle-iron supports to accommodate the weight of the column. Installation of the column poses little problem, for it can be packed after it is set upon the boiler. Removal for access to the boiler's interior is another matter. You may prefer permanent installation of the column and fabrication of a small door (hatch) on the boiler for easy access.

How to Build a Packed Column Still. Wood is by far the easiest material to work with. One 4′ × 8′ × ¾″ sheet of marine plywood will give you a 1-foot-square, 8-foot-tall column, relatively weather resistant. Prior to construction, coat all surfaces–sides and edges–with two coats of alcohol-base paint.

Use a straight edge–a two-by-four clamped to the plywood–when cutting the 1′ × 8′ strips. Any irregularity or waviness in the cut will create gaps upon fabrication, which may allow alcohol vapor to escape during distillation.

Just prior to joining the four pieces, seal all butting edges with a bead of silicone or high-temperature glue to prevent vapor escape during distillation.

Square collars can be used to shape and hold your column. Place one every 2 feet along the length. These can be permanently affixed for added structural strength. Screws are preferred throughout fabrication for holding and staying power. Contraction/expansion during use will eventually loosen nails.

A pack-material retainer must be installed near the bottom of the column. Hardware cloth (screen) ¼-inch mesh, with edges bent back 2 inches to form an inverted basket will work well. Use two layers for strength and durability, and attach with screws to the interior walls of the column.

Plumbing and adaptors can be added to the top of the column now or after the column has been installed on the boiler. (See Plate Column, Final Stage Condensation.)

Sheet metal may also be used to construct the packed column still, although it is considerably more involved and complicated. The advantage to using metal is longevity. However, this can be somewhat offset since wood is so much easier to replace when it finally becomes necessary.

An option to making your own metal column is to buy heating-system duct work. Relatively inexpensive, it is easy to work with and will do the job you need done. Although you must still assemble and adapt component parts, you save the bother of welding and/or joining side seams. Cost and personal preference will be the two determining factors here.

The Plate Column Still. If you make the column shell of wood, it will take longer than making a packed column, but it will not be that much more difficult. If it is made of metal, you will have a great deal of fitting and welding to do.

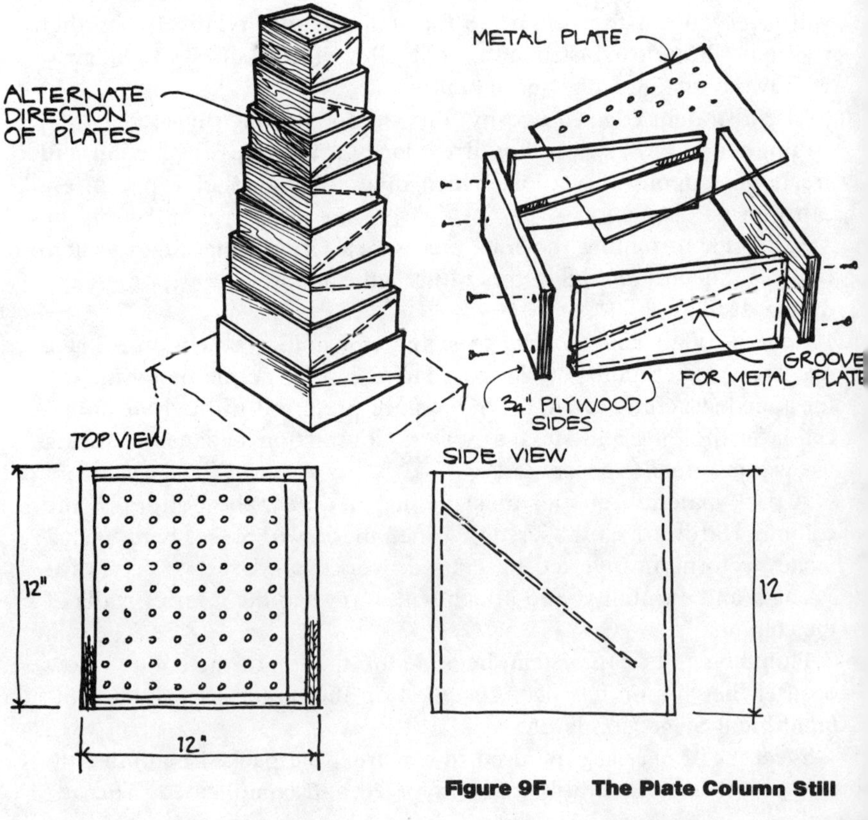

Figure 9F. The Plate Column Still

The plate column works on the same principles as the packed column—173°F ambient temperature, and surface-area volume. However, instead of packing material into the column, flat metal plates are installed for initial and intermediate condensation. Placed at a 30- to 45-degree angle every 12 inches throughout the column, the plates condense water vapor and let the alcohol pass through to the top.

Although solid metal plates could be used, sieve plates (holes pierced throughout) are more efficient in that they allow alcohol vapor to rush up the column more easily and readily. A sieve plate should have one hole per square inch.

How to Build a Plate Column Still. When complete, a wooden plate-column still will resemble a four-sided, stepped pyramid, with vertical sides. Beginning with a 2-foot-square base section, each subsequent section fits into the preceding section for a progressive size reduction of 1½ inches per section. The top section on an 8-foot-tall column will be 1′ × 1′ square. (See page 116.)

The plates can be from ⅛ inch to ¼ inch thick. The smaller size is easier to work with—cut and drill—and fit to the interior walls of the column. The larger offers longer heat retention, which is beneficial if you plan to make multiple and consecutive distillation runs.

Begin by cutting four strips of ¾-inch marine plywood—two pieces 12″ × 24″, and two pieces 12″ × 22½″. When assembled, exterior measurement will be 24″ × 24″ and the interior, 22½″ × 22½″. Apply two coats of alcohol-base paint. Cut one slot in each piece of one set—either set will do—¼ inch deep and 22 inches long to receive the metal plate and hold it in place. The width of the slot will depend upon the thickness of the metal plate. This slot should begin at the upper left-hand corner, 2 inches down from the top and ¾ inch in from the side, and run down to the lower right-hand corner 2 inches up from the bottom and in from the side. The slot on the other piece will start at the upper right-hand corner and run down to the lower left-hand corner using the same dimensions as the first piece. When the metal plate is installed, it must rest against the side wall perpendicular to these two pieces. This prevents water vapor from sneaking past the plate without contact and condensation.

The dimensions of this first plate will be 22½ inches plus ½ inch (to accommodate the ¼-inch slot in each side piece), or 23 inches, by 22 inches. Final assembly is done by inserting the metal plate into the slots and fastening all sides with silicone and screws.

The second section will be slightly smaller than the first and have the following dimensions: exterior–22½″ × 22½″; interior–21″ × 21″. The metal plate will be 22″ × 21″. Use the same construction and assembly as above. When assembled, this second section will snugly fit within the first section and rest on a ¾-inch shelf placed 1 inch down from the top of the first section. Fasten the two sections with four screws, one at the midpoint on each of the four sides. Seal the units with a bead of silicone around the entire perimeter.

Continue making the sections as above, progressively smaller, until you have eight completed. The last section will have the dimensions 12″ × 12″, with a plate 12½″ × 12″.

NOTE: When installing the sections, the plates should alternate 180 degrees, running down left to right, then down right to left, and so on. This forces vapor contact with metal at each section and increases efficiency.

With a delicate hand and a little patience, you can avoid section construction by building three sides, 8′ × 1′ each side, with two of the sides slotted; inserting the metal plates, alternating 180 degrees each foot; and installing the fourth side, sealing the unit.

If you build your plate column still out of metal, the sections are constructed basically as above, except that dimensions and fastening differ. All plates and most seams are welded, and the column need not be step pyramided–it can begin with a 1-foot-diameter base and end with a 1-foot-diameter top. Sections can be welded together as they are placed, or flanges can be welded to each section–top and bottom–and the sections subsequently bolted together. However, flanging requires a gasket, which may eventually deteriorate, cause leaks, and need replacement.

Three-sided fabrication and placing of plates is possible, though tricky. The plates must be alternated, so one side of half of the plates cannot be welded to the fourth side of the column wall when it is welded in place. If you are a welding novice, take the time and build it in sections.

The column can be attached to the boiler in the same manner as the cap. A flange that will snugly receive the column must be fabricated onto the boiler top, and for a more secure fitting, bolts can extend out from the flange to mate with elongated slots cut into the column sides. Wing nuts then tighten down to draw the flange and the column sides together. Silicone can then be run around the perimeter to effectively seal against leaks during operation.

The top of your column can be topped off with a square-to-round adaptor and size-reduction ductwork; or, if made of wood, a 4-inch round heating duct or vent pipe can be fabricated to a 1-foot-square piece of wood, which is subsequently attached to the top of the column. Regardless of choice of the above two, a right-angle elbow is attached to begin piping away from the column. To this is attached a 4- to 5-foot length of 4-inch vent pipe that begins alcohol vapor condensation and leads to final-stage condensation.

Final-stage condensation can occur in an air-cooled copper coil or a rectifier. A rectifier is of basic column design. However, the sieve plates are installed horizontally (flat) instead of at an angle. The alcohol vapors enter from the top and are forced down over the plates, condensing as they drop.

Although metal side-wall construction is preferred for heat dissipation, wood can be used. The sieve plates are placed every 6 inches for maximum surface-area contact with vapors. The top must be sealed (with silicone) to prevent vapor loss. Condensed alcohol collects on the bottom of the column, so naturally a drain pipe is necessary for collection of the finished product.

COLLECTION AND DENATURING

Collection. Regardless of the type of the still you use, a small amount of alcohol vapor is bound to escape during collection. The preferred collection container is therefore made of metal, which will inhibit evaporation and loss of vapor. A narrow-mouth container should be used to also diminish the loss of vapor. Glass containers should be avoided because of the cracking and breaking potential and poor heat-dispersal qualities. Permanent storage should be in an approved vented tank.

Denaturing. Federal law mandates the addition to every one hundred gallons of alcohol one gallon of either gasoline or kerosene. This is to be done at the point and time of collection. This is simple enough: If you are using a hundred-gallon collection tank, add one gallon of gasoline or kerosene to the collection tank prior to firing your boiler. To a fifty-gallon tank, add one half gallon, and so on. Although either gasoline or kerosene can be used, kerosene is preferred mostly for the sake of cost–it will always be less than the cost of gasoline.

EFFLUENT

Once distillation is complete, you can either reuse the effluent for fermentation or dispose of it. Effluent disposal can best be facilitated by using a septic tank principle. Dig a rectangular pit (close to your distillation equipment) with the same dimensions as your boiler. Fill this pit with loose gravel, and as you complete a distillation run, empty the effluent into the pit. Most of the effluent will be absorbed by the earthen walls of the pit, while some will be gradually lost to evaporation.

Certain crops grow best within certain pH ranges. If your soil is alkaline, and your particular crop likes acidic soil, you can spray (or pour) the acidic fluid effluent onto the soil. Doing this will help adjust the soil's pH to a more favorable crop-growing condition. If your soil is already acidic enough, you may choose to allocate a small portion of your land for surface-absorption effluent disposal, instead of digging and installing a septic-type system. Here, you would simply dump your effluent onto flat soil (furrowed or graded smooth) and let it soak in. Sloped or hilly terrain is not that great, for erosion will eventually occur.

HOW TO RAISE THE PROOF OF YOUR FINAL PRODUCT

A typical moonshiner still will produce from 160 to 175 proof alcohol; a column still, from 175 to 190 proof; each depending upon construction variables. There is a simple enough way to raise the proof of your final product to 195–196 proof without distillation. This means a better product without the extra fuel cost of firing a boiler.

The process is basic filtration and works on the principle that the specific gravity (density) of water (1.0) and of alcohol (.79) are different. The higher-density water is trapped by the filtering agent and held, while the lower-density alcohol slips on through. This is primarily a mechanical process; there is little or no chemical reaction involved, although some filtering agents may cause a small reaction with either the alcohol or the water. The most common filtering agents are grain and zeolite (aluminum silicates: sodium, calcium, potassium, or barium). Grain is the most readily accessible to the average person, and may be incorporated into your system rather easily.

The deterrent to filtration is volume. If you are processing only several gallons of alcohol, forget filtration, for it takes several gallons just to charge your system. A bushel of grain—say, cracked corn—will absorb about three to four gallons of fluid before reaching saturation. Since you are concerned only with about 10 percent by volume of water, one bushel of corn will do quite a bit of filtering before reaching total saturation, i.e., before water begins coming out with the alcohol. Exact quantities depend upon the initial proof of the alcohol fed into the system. A higher-proof alcohol will permit more gallons to be filtered since less water is being absorbed.

Any container with an open top and a drain on the bottom will work as a filtering column. The primary considerations are capacity (which depends upon the volume of your operation), ease of removing the filtering agent, and the type of filtering agent (zeolite or grain).

The disadvantage of zeolite is that the captured water must be removed either by heating the column or by running hot gas through the column after filtration. The captured water will not dissipate by itself. This purging requires extra time, work, and fuel, which increases the complexity and cost of your operation. The advantage of zeolite is efficiency; it is much more efficient than grain and is good for several thousand runs before replacement becomes necessary. Grain must be removed from the column after two to three days to avoid spoilage or premature fermentation. Of course the grain can be used for mashing and fermentation after removal from the column, so you really only incur an extra handling step.

A potential problem with grain filtration is drain/valve clogging. As the grain becomes moist, starchy lumps form, which can inhibit the free flow of alcohol. To avoid this problem, you need a layer of chunk-style material between the drain and the grain. Regardless of the drain's diameter, line the bottom of your container with ¼-inch hardware cloth (screen) over which you lay 2 inches of coarse gravel, over which you lay 2 inches of charcoal (barbecue grill type will do). This prevents drain clogging with the added benefit of the charcoal filtering out grain particles that might otherwise have made it past the gravel and to the drain. When the system is purged, the charcoal and grain are removed together, the charcoal separated and air dried, and then used for either another filtration cycle or for a barbecue.

For continuous multiple runs over a two- or three-day period, a hydrometer will tell you when the grain has reached saturation. When the proof coming out is less than the proof you put in the top, the grain has passed its usefulness as a filter.

WHAT FUEL SOURCE SHOULD YOU USE?

There are two basic questions you must ask yourself before deciding on a fuel source. Even then, cost will be a major determining factor.

1. Do you want a heat source that requires installing plumbing, fixtures, valves, and pumps, after which you light it and literally walk away from it?
2. Do you want a heat source that takes some effort to light and must thereafter be tended?

The first option requires an initial investment for material and labor. Even if you install it yourself, your labor value must be considered. Fuel oil, kerosene, natural gas, and propane gas all require some form of fuel delivery and fixtures for spraying or injecting the fuel for ignition. Propane and natural gas are by far the easiest to implement—all you need is a line (piping) for fuel delivery to the boiler bottom, a valve to regulate the flow of gas, and a match to light it. Fuel oil and kerosene are a bit more complicated. However, once you have a few distillation runs timed, you can just about light it and walk away from it for an hour or so. The major disadvantage to these sources is that you may install your equipment and then experience a drastic increase in fuel cost. Conversion cost must be considered.

The second option depends upon availability and accessibility, and storage facilities. If you can cut and store your own wood at little or no cost, then wood is most likely the fuel source for you. If you must buy wood from a commercial vendor, cost takes precedence. Coal is a viable fuel source if you have storage space and the cost is right. With both, it takes time to get a good fire going, and both must be tended periodically—fuel added, ashes and hot coals shuffled around.

Newspaper logs—made from wet newspaper rolled, tied, and left to dry—are just about as good as wood logs as a fuel source. The ideal feature of newspaper logs is that if you make weekly neighborhood collections, your source is literally free. Most people throw their daily paper away, so let everyone know that you are interested and that you will pick up regularly. Don't forget magazines. Most people throw them away also. If nothing else, magazines are good fire starters. Population density has a major effort on whether using newspapers as a fuel source is viable. If your closest neighbor is two miles away, it is hardly worth scouring the countryside for a handful of papers and magazines.

Segregating your personal garbage—flammable from nonflammable—will not provide a primary fuel source, but it will help a little. Trash compactors in your neighborhood? Solicit their bundles of squeezed-together trash, whether or not the trash has been segregated. If it contains nonflammable material, you can always bundle it up for disposal, along with your regular garbage, after a distillation run. (Caution: Avoid aerosol cans that will explode when heated.)

Commercial and community waste. The questions here are: Is it wet or dry, and does it cost you anything (other than pickup time and effort)? If the waste is wet, it will have to be dried before use as a fuel, which is probably not worth the effort. If damp, sun drying may do the job. If dry and free, consider it as a primary fuel source.

DETERMINING COST VERSUS BTU

The cost of various fuels will always vary, and it is somewhat confusing to compute and compare fuels because they are not all measured by the same standard of volume. Coal is generally measured in tons (cost per ton), wood is usually sold by the cord, fuel oil and kerosene by the gallon, natural gas by the cubic foot, and propane gas by the pound or gallon. For cost-comparison accuracy, all fuels considered must be gauged alike—by the heat value or Btu factor of each fuel. You can use the following chart to help in your comparison.

Btu Comparison Chart

Fuel Stock	*Btu Rating per Unit Measure*
Agricultural residue	3,000 to 8,000 per pound
Coal: Lignite (extra soft)	7,000 per pound
Bituminous (soft)	13,000 per pound
Anthracite (hard)	13,000 per pound
Wood	7,500 to 8,700 per pound
Fuel oil	140,000 per gallon
Kerosene	90,000 per gallon

Btu Comparison Chart

Fuel Stock	*Btu Rating per Unit Measure*
Commercial/community waste	8,000 per pound
Natural gas	1020 per cubic foot
Propane gas	2690 per cubic foot

Your primary concern in cost comparing is finding out the cost per 100,000 Btu of any given fuel. This means that in some instances you will be buying by the pound, by the gallon, or by the cubic foot. As you contact distributors, you simply ask the cost of 100,000 Btu and fill it in on the chart below. If, for whatever reason, you are quoted a per-unit-measure cost (for example, cost for a ton of coal), simple arithmetic will give you the cost for 100,000 Btu based on the chart above.

Cost Comparison Chart

Fuel Stock	*Cost per 100,000 Btu*
Agricultural residue	$________
Coal: Lignite (extra soft)	$________
Bituminous (soft)	$________
Anthracite (hard)	$________
Wood	$________
Fuel oil	$________
Kerosene	$________
Commercial/community waste	$________
Natural gas	$________
Propane gas	$________

THE SOLAR STILL: WHAT IT'S MADE OF AND HOW IT WORKS

A solar still works on the same principle as solar panels used to heat a house or hot water system: heat absorption and retention. However, instead of removing the heat from the unit for distribution by hot-air convection or metal coils (as with solar panels), the goal with a solarized still is to absorb and retain as much heat as possible within the unit.

The basic construction consists of a waterproof wood or metal box with either a glass or plastic top. The box holds the beer; the glass/plastic top admits sunlight to heat the beer, and to capture and condense the resulting alcohol vapors.

The major advantage of a solar still is that your heat source is free. The major disadvantage is that the final proof alcohol on one solar still run is only 120 proof. Successive runs will naturally increase the proof, but it takes twenty to twenty-five individual runs to match a single run using either a moonshine still or column, respectively. It is also difficult, without using some arithmetic, to tell when a solar run is finished. The ambient temperature does not reach 173°F–distillation is primarily accelerated evaporation–so there is no distinct cutoff when alcohol stops flowing and water starts, nor is there a distinct and sudden increase in temperature to mark conclusion.

So why even consider a solar still? Because the heat source is free, and because it can be used in conjunction with a heat-fired still to literally cut your fuel cost in half–if not more, depending upon the exact proof of the alcohol involved. This is the difference between the energy needed to heat and distill 90 percent water and 10 percent alcohol, and that necessary to distill 60 percent alcohol and 40 percent water. The total Btu necessary to raise the 60/40 blend to 173°F is less than is needed for the 90/10 because the specific gravity of water is about 21 percent more than that of alcohol. Of course, the heat-fired still must run longer to fully distill the 60 percent alcohol, but the total yield will be about five to six times greater by volume than an equivalent initial volume of 90/10.

HOW TO BUILD A SOLAR STILL

The basic box can be of wood or metal, although wood is preferred for ease of fabrication and for heat retention. Metal, unless well insulated, will cool the fluid just about as fast as the sun heats it.

The total volume of the still should be at least three times the volume of beer you plan to use on one run. The still should be a shallow, rectangular box with a large top surface area of glass or plastic. For example, if you plan a 500-gallon run, the conversion of gallons to a base volume in cubic feet is about 60 cubic feet (500 divided by 7.48).

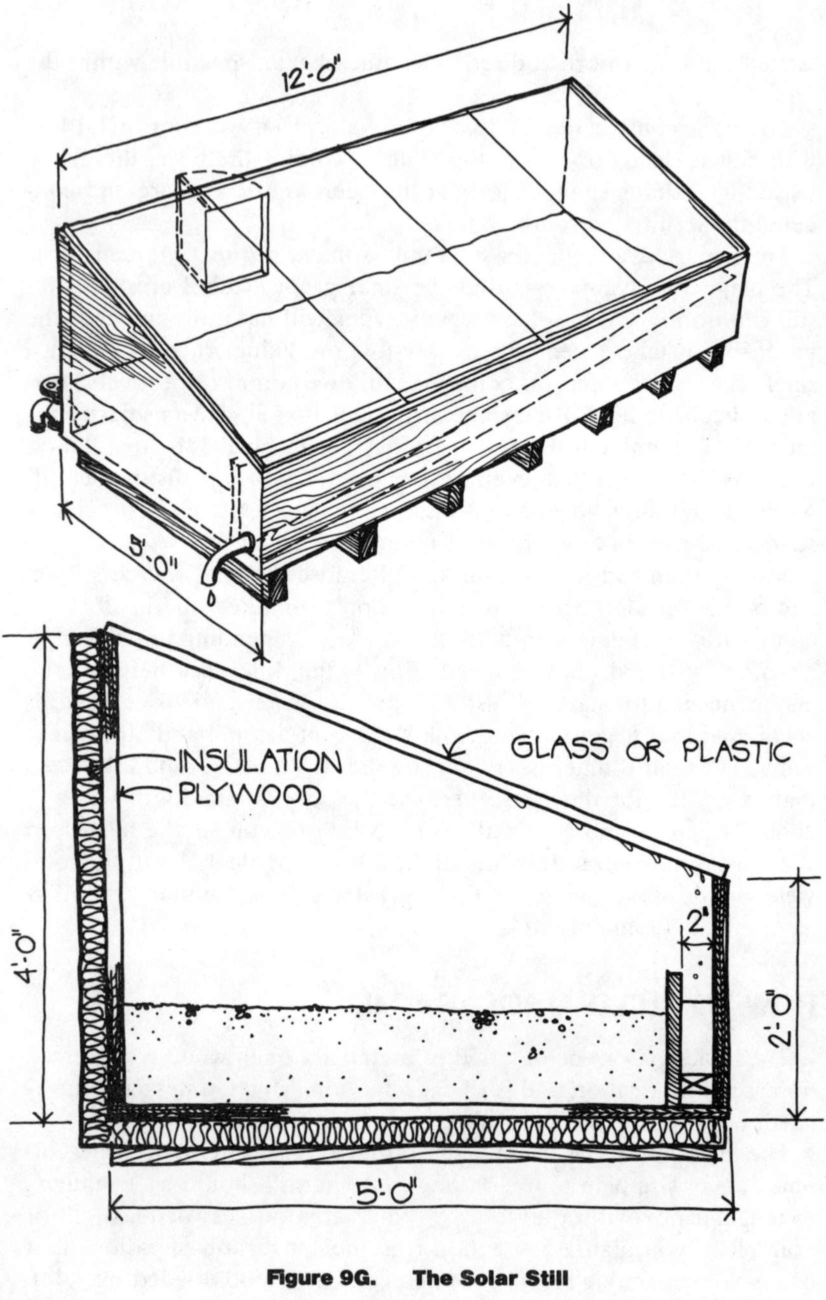

Figure 9G. The Solar Still

A 5′ × 12′ base will accommodate 500 gallons of fluid 1 foot deep. The remainder of the still should then be at least 120 cubic feet to facilitate condensation of vapor and collection of alcohol. Since the side walls must be tapered to give the glass the desired 30- to 45-degree angle when installed, a little simple arithmetic is involved. Without using complicated algebra or geometry, you can come close to the desired measurements using a stacking method of computation. (See Figure 9H.)

A 5′ × 12′ × 1′ box gives you 60 cubic feet. Construct—either on paper or mentally—a rectangular frame 5′ × 12′ × 1′ and you have another 60 cubic feet. Bisect this resultant 5′ × 12′ × 2′ box by drawing a line from the top left corner to the bottom right corner, which will yield two wedges of 60 cubic feet each. Place one wedge on top of the 5′ × 12′ × 1′ box and you have a total of 120 cubic feet with side dimensions of: width, 5′; back wall height 4′; front wall height, 2′; length 12′.

Basic construction is similar to building a fermentation tank or wood column still, although it is advisable to use a frame of two-by-fours to which ½-inch plywood is laminated. When cutting the sides,

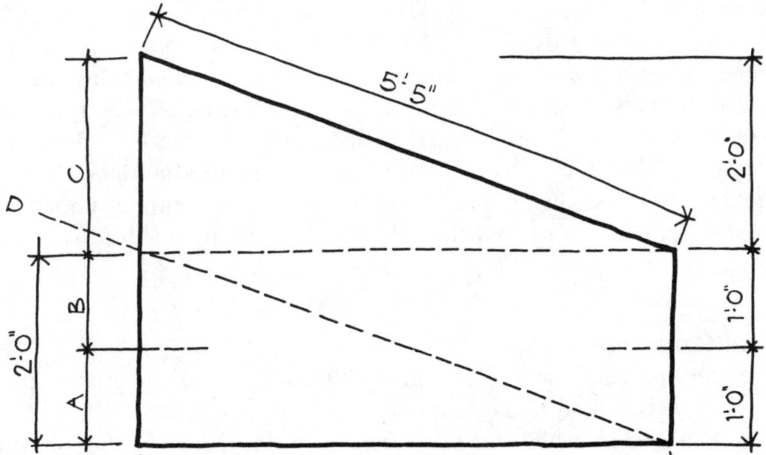

Figure 9H. Construction of the Solar Still Base Box

clamp them together for one cut so that the angles are identical. Bevel the top of the back so the angle accommodates the glass (plastic) when installed. To all top surfaces that will be in contact with the glass (plastic), place a bead of liquid silicone and flatten it with a putty knife to a uniform thickness of ⅛ inch to ¼ inch. This forms a seal between wood and glass to prevent the loss of vapor during distillation. If gaps are evident when the glass is installed, remove and add more silicone where necessary so you have a hundred-percent surface contact between glass and silicone.

The front panel should extend ½ inch above the intersection plane of the side panels against which the glass panels can rest. This prevents the glass from sliding down, and facilitates condensation collecton. As the droplets run down the glass, they hit the front wall and run down into the trough. Obviously, a retaining panel must be installed within the still to separate the beer from the collection area. This should be 18″ × 12′ and installed vertically 2 inches from the front wall, running the full length of the still. This should be sealed, both sides and bottom, to prevent beer leakage into the collection chamber and contamination of the final product.

Collection Chamber. Condensation and collection of product in a solar still is strictly by gravity. The basic consideration, then, is that once vapor condenses, the droplets run down to a final small point of collection. This means installing a 2″ × 12′ strip of wood in the chamber that runs from a 2- to 3-inch height from the base down to base level. This, too, must be sealed to prevent product loss. As the alcohol runs down the front wall, it hits the angled chamber floor and continues a downward path to a 1- to 2-inch-diameter drain/collection pipe.

Final Considerations.

1. During construction, seal all joints with silicone or glue to prevent leaks and loss of vapor.
2. Once constructed, all surfaces (excluding glass/plastic) should be coated with two coats of a heat- and alcohol-resistant flat black paint. Do not use glossy black—flat absorbs heat much better.
3. Two drains are absolutely necessary: one for alcohol collection, and the other to purge the spent beer after distillation is com-

plete. Charging the still can be done through 2- to 3-inch piping, or via a small hinged access door on the back wall. A door will require a rubber gasket to prevent vapor loss, and, of course, a latch to securely close it.

4. For economy, ¼-inch plywood can be used instead of ½-inch, provided you use frame construction for structural strength.
5. Glass or plastic panels can be any width provided they run top to bottom without interrupting condensation flow. For the described solar still, you could use twelve 1′ × 5′5″ panels placed side by side across the full length of the still. The cost of one 5′5″ × 12′ glass or plastic panel is extremely high. The weight and breakage potential make handling difficult. Small panels are easy to handle and relatively inexpensive, and breaking one panel is nothing compared to breaking an entire sheet. If you choose to use panels, you must install a top-to-bottom support between each and every pair of panels. These supports must also be coated with silicone for sealing purposes.
6. The larger the volume of your still, the heavier it will be, and consequently the more difficult to move. Therefore, before you construct it, decide its final resting place and build it on the spot. It must be elevated from the ground–2 inches will do–to prevent eventual wood rotting (or metal rusting). The front of the still must face south, with the side walls facing east and west respectively for maximum exposure to sunlight.

Options.

1. Instead of painting the interior of the still, you can line the bottom and side walls with corrugated metal painted flat black. The metal will absorb and retain more heat than wood will, and will decrease total still operating time considerably. Small clips can be used to hold them in place.
2. Painting the back wall white is effective for reflecting sunlight onto the fluid, but far more effective is laminating scrap mirror onto the wall. Mirror shops generally discard small scraps that cannot be utilized commercially. If you have access to enough pieces, glue them to the side walls down to the fluid-level line for maximum reflection.
3. If you care to go one step further, you can erect a flat or slightly concave makeshift aluminum reflector around the perimeter of the still, angled to direct sunlight into the still.

SOLAR STILL FLOW PROCESS

Once the still is constructed and in place, solar distillation is relatively simple. Charge the still with beer, smear a thin layer of axle grease (about ⅛ inch) on all edges and surfaces that will come in contact with glass/plastic for additional seal, clean glass panels and put them in place, and take a break. The still and the sun will take it from there and you really need only check the collection end periodically to gauge distillation progress and conclusion.

How to Know When to Stop. Solar distillation is almost foolproof. If there are no leaks anywhere, you should—depending on the ambient temperature—begin to condense alcohol within the hour of starting. The trick is to know when to stop: when have you reached the highest proof possible. This can be done with a hydrometer or with arithmetic. If you use a hydrometer, check the condensation for proof level when you have about a quart of fluid. Jot that down. Thereafter, check proof periodically—about every half hour or so—and when the proof begins to drop, distillation is concluded. After several runs, provided there are no major variables (you do not change mash formula and are using the same quantity to charge the still), you will know how long it takes a given run at a given temperature. A 95°F day will naturally work faster than a 75°F day, so bear temperature in mind. If you use arithmetic, all you need watch is the total volume produced. If your hydrometer indicates that your beer is 10 percent alcohol and you are making a 500-gallon run, then complete distillation would yield 50 gallons of alcohol (500 × .10). It is safe to presume a final solar yield of 120 proof, since during evaporation 6 parts alcohol and 4 parts water will simultaneously leave the fluid's surface. Since your end product will be 60 percent alcohol and 40 percent water, you then divide 50 gallons by .60 (60 percent) to get a total volume yield of 83.3 gallons. Therefore, on a 500-gallon run, you would conclude distillation when you have collected about 83 gallons of fluid.

You can now immediately distill this 120 proof alcohol in your heat-fired still, or you can store it until you have accumulated enough to fill your still. This really depends upon the capacity of your still. If you have a 500-gallon still, you can save four or five solar still runs and process them all at one time.

Solar Still Dos and Don'ts

1. Don't start a solar still run too late in the day. Any time up to 11 A.M. is acceptable. If you start too late in the day, you will lose optimum sunlight and heat.
2. Don't place your still where it will be shaded at any time of the day, such as in line with a building, trees, etc. Wide-open space is best.
3. Don't collect your product in a wide-mouth container. Solar distillation is a very slow process. Conceivably, you can lose your product just about as fast as it is made, by evaporation from the collection container.
4. Do clean everything after a solar run (a mild bleach solution is ideal): Flush the tank out completely with fresh water, wash the panels until they sparkle, and check all components for wear and tear. If you use corrugated metal as a liner, remove and clean after each run. The tank must be cleaned to prevent residue buildup, and to eliminate potential bacterial breeding. Panels must be clean to allow maximum sunlight penetration on the next run.
5. Do not use any commercial cleansers or solvents that may leave a residue.

DISTILLATION DOS AND DON'TS

1. Do not fill your boiler to its absolute top capacity. Leave a little room for bubbling and spitting. This is especially true if you are using a column still.
2. Do not fire an empty boiler. Although it may seem to give you a head start, when cold beer hits hot metal it will do two things: buckle and warp your boiler; and generate immediate steam (water vapor), some of which will escape initial condensation (cap or column).
3. Once you start a run, do not touch anything–cap, column, plumbing, etc. Any and all adjustments should be done prior to firing the boiler.
4. If anything goes wrong or needs adjusting during distillation, shut down and make the necessary corrections. Take absolutely no chances.

5. Cleanliness counts. Residue buildup decreases efficiency, and increases corrosion potential. The cleaner your equipment, the longer it will last.
6. Check all seams or welds periodically for distress, leaks, and signs of wear. Once a seam starts to go bad at any given point, it can rupture the full length at any time without too much warning.

WHAT TO CONSIDER IF YOU CHOOSE TO BUY A STILL

If you plan to buy a still instead of building your own, there are certain questions you should ask and certain considerations that should be made. The following guidelines will give you some basics, but more important, ask a lot of questions and do a lot of comparison shopping before you finally buy.

1. What is the type and thickness (gauge) of metal used for still construction? Two stills may cost the same and look the same, but the thinner gauge will not last as long as the thicker. Of the types of metal used, the following are preferred, in this order: copper, stainless steel, cast iron, carbon steel, aluminum.
2. What safety features are offered? Blowout valve? Temperature and pressure gauges? Automatic shutdown controls?
3. What is the cost per gallon of alcohol per hour of operation per capacity?
4. Warranty or guarantee. What is covered for how many years? Most important, what is not covered? What is the replacement cost for parts not covered?
5. Ease of operation. How much is automatic and how much requires manual labor? Is the equipment easy to clean and maintain? Are critical parts accessible?
6. How available are parts–maintenance and replacement–and how expensive?
7. How does the general construction compare with others for durability and longevity?
8. What is the gallons-per-hour rate and proof of the final product? This is an efficiency rating. If the final proof is only 120, the still is not that efficient. You could do as well with a homemade solar still.

9. What is the Btu input for a single run and how does it compare with others?
10. What is the heat source (gas, electric, oil) and is it built in to the design? And how easy is conversion from one fuel to another?
11. Dealer. How far is he from your operation and how is his reputation?

10
Using What You Have Made to Fuel Your Car or Truck

GASOHOL

GASOHOL IS the most readily applied use of the alcohol you have made. It is popularly defined as a mixture of anhydrous alcohol and gasoline in a proportion of 10 percent alcohol to 90 percent gasoline. A higher proportion of alcohol can be added and it will still be called gasohol. A blend of up to 30 percent alcohol/70 percent gasoline can be used to fuel an internal combustion engine—your car for example—with no engine (carburetor) modifications.

Gasohol is made by blending 195- to 198-proof alcohol (ethanol or methanol) with unleaded gasoline. The resultant blend will have an antiknock (octane) rating one point higher for each 3.5 percent alcohol added. Ten-percent gasohol (contains 10 percent alcohol) will have an octane rating three points higher than the gasoline into which the alcohol is placed. This basically means that a lower-octane gasoline (cheaper to produce, refine, and buy) can be used to manufacture commercially pumped gasohol, which will subsequently make the gasohol price competitive with that of unleaded gasoline. As it is, gasohol currently costs a few cents more than gasoline. For the do-it-yourselfer, this means that you can buy cheaper gasoline that would ordinarily not burn well in your car and make your own gasohol that *will* burn well.

PROBLEMS TO CONSIDER

Water is the greatest problem associated with blending, storing, and using gasohol. Absolute anhydrous alcohol (200 proof) does not remain absolute very long. Exposed to air, it immediately absorbs moisture from the air and the proof instantly drops. How much of a drop depends upon the ambient temperature, humidity, and atmospheric pressure. Eventually, the proof will drop to about 195–196 and stabilize. When this is mixed with gasoline, the moisture settles out from the gasohol blend and, being heavier than gasoline, lays on the bottom of the tank—storage tank, or your car's fuel tank.

Even mixing alcohol and gasoline in a settling tank and removing any water that settles out does not assure problem-free gasohol. Gasoline itself will render a few drops of water as it sits, depending on the period of time it sits and on the temperature. When alcohol is added to gasoline, it does two things: It attracts moisture, which will not mix with gasoline and therefore settles out; and to some degree it accelerates gasoline's natural tendency to render water. Since your engine will not run on water, this presents some very obvious problems.

If gasohol is blended before delivery to a service station (distributor), the only problem you may encounter is if atmospheric water has settled in the station's pumps and is pumped into your fuel tank. It is the station's responsibility not to pump water into your tank, and the station's liability if it does so, creating problems for you.

If the station mixes the alcohol and gasoline as it is pumped into your car, the proof of the alcohol must be 199 to preclude problems of water settling out after the newly blended gasohol enters your fuel tank. Since it is extremely difficult to store alcohol and retain the 199 proof status, your best defense is to buy only preblended gasohol, i.e., not blended at the point of purchase.

Vapor lock is a potential gasohol use problem depending on your particular car and its fuel delivery system. Vapor lock usually occurs between the fuel tank and the fuel pump. It is, in a sense, a vapor bubble in the fuel line which interrupts the normal constant flow of fuel. It is caused by heat—either a very hot summer day, or a fuel line that runs too close to an exhaust manifold.

If you have ever experienced vapor lock with your car using regular gasoline, you can expect the same problem more frequently with gasohol since alcohol increases the vapor pressure (or vaporization) of gasoline.

If your fuel line does not run near an exhaust manifold, and you have ever had vapor lock, there is little you can do to avoid future occurrences. However, if the fuel line is close to a heat source, you can install a heat shield or deflector to minimize the chances of vapor lock.

If a heat shield is not viable, wrap the line with insultape or other insulating material.

HOW TO MAKE YOUR OWN GASOHOL

Adding the required 1 percent gasoline (or kerosene) to your alcohol to denature it is no problem regardless of the size of your operation and regardless of the final proof of your alcohol. However, adding 10 to 30 percent alcohol to unleaded gasoline to make gasohol is another matter. Since most small-scale stills are not designed, nor efficient enough, to produce 199 proof alcohol, you must add water-alcohol to gasoline, and then bleed off the resultant water that settles out of solution. You cannot, *absolutely not*, just add the alcohol to your car's fuel tank and hope to run your car for more than a mile or two. As soon as the water, which will settle to the bottom of your tank, reaches your carburetor, your engine will suffer a massive H_2O embolism. It will not run, and it will not start.

To make gasohol from 120- to 190-proof alcohol, it must be mixed in a separate tank, preferably immediately following distillation while the alcohol is still warm. The amount of alcohol received by the gasoline and amount retained by the water that settles out will vary depending on the temperature of your fluids, agitation (stirring), and alcohol proof.

The higher the temperature of the alcohol, the more readily the alcohol will leave behind the water and blend (bond) with the gasoline. Flash points must be observed here for safety's sake. Alcohol at a temperature between 125°F and 150°F is relatively safe provided there is no flame or spark nearby. Caution is necessary if electrical switches or outlets are in proximity to your blending operation. Having adequate ventilation is the best way to avoid problems.

Agitation or stirring increases the number of gasoline molecules in contact with alcohol molecules. While not absolutely necessary, stirring expedites blending.

The higher the alcohol proof, the more alcohol that will go into solution with gasoline. For example (and this is approximate), if you mix 160 proof (80 percent) alcohol with gasoline, the water will take about 20 percent of the alcohol with it as it settles out, so that, by volume, only 60 percent of the alcohol mixes with the gasoline. With 180 proof, the water takes about 10 percent of the alcohol with it and 80 percent mixes with the gasoline.

This is significant for two reasons: It increases your process time, for you must redistill the water-alcohol that settles out or waste the alcohol; and it affects the volume ratios of how much water-alcohol must be mixed with gasoline to achieve a certain-percent gasohol.

Working on a percent basis, to make 10-percent gasohol you must mix 1 gallon of 199 proof alcohol with 9 gallons of unleaded gasoline. However, you would need to add 1.1 gallons of 190 proof alcohol to 9 gallons of gasoline to get a 10-percent gasohol blend, 2.2 gallons to 8 gallons for a 20-percent blend, and 3.3 gallons to 7 gallons for a 30-percent blend. The following table will simplify your blending ratios.

PROOF	*10% GASOHOL* To 9 gals. gasoline, add:	*20% GASOHOL* To 8 gals. gasoline, add:	*30% GASOHOL* To 7 gals. gasoline, add:
190	1.1 gals.	2.2 gals.	3.3 gals.
180	1.2 gals.	2.4 gals.	3.6 gals.
170	1.4 gals.	2.8 gals.	4.2 gals.
160	1.7 gals.	3.4 gals.	5.1 gals.
150	2.0 gals.	4.0 gals.	6.0 gals.
140	2.5 gals.	5.0 gals.	7.5 gals.
130	3.0 gals.	6.0 gals.	9.0 gals.
120	5.0 gals.	10.0 gals.	15.0 gals.

It is neither practical nor advisable to attempt to make gasohol with less than 120-proof alcohol. Most of the alcohol will settle out, leaving you not much better off than when you started—wasting time and effort.

How to Determine the Optimum Percent Blend for Your Car. How high a percentage of alcohol should you use—10 percent, 20 percent, 30 percent, or more? This depends on your particular car, and the time of year—summer or winter. In the summer, higher temperatures increase the water absorption potential of your gasohol depending on the percent alcohol present—the more alcohol, the more water absorption potential. In the winter, vaporization becomes a problem. A higher percentage of alcohol creates cold morning starting problems, engine hesitation, poor acceleration, and general deficient performance. The seasonal effect really depends upon the ambient temperature—a cool summer or mild winter will naturally allow you to use a higher percentage. The ideal temperature range for 10–30-percent gasohol is 55°F to 65°F.

The condition of your engine has a major bearing on performance. If you have any problems at all running your car on straight gasoline, those problems will be compounded by gasohol usage. Your engine must be in top time and tune condition before you can successfully use any percent blend of gasohol. The ideal percentage blend of gasohol for your car is best determined by starting with a 10-percent blend and progressively increasing the percentage until you experience problems—poor starting or performance. At the point that problems develop, deduct 5 percent from what you are using and stick with that. Generally, you will find 20 to 30 percent to be the maximum limit without having to make engine modifications. A 30- to 40-percent blend usually requires minor idle screw and fuel ratio screw adjustments. At a blend of about 50 percent, the timing must be advanced, using a timing light, and carburetor adjustments must be made. Above 50 percent alcohol, the carburetor fuel jets may need to be enlarged. (See Chapter Eleven for adjustment variables.)

Separation Tank. Any container with a drain valve on the bottom will serve well as a separation tank. A clean, fifty-five-gallon metal drum is ideal, offering both capacity and convenience. A fluid level sight-glass (see Fig. 10A) is not mandatory, although it does let you know the level cutoff between gasohol and water. A closed top with funnel is preferred over an open top, to avoid loss of fuel to evaporation. There are two options for adding your alcohol to the unleaded gasoline: Add it from already-bottled and stored alcohol of a known proof, or add it as it is manufactured right from the condenser into the separation tank.

Option One. The alcohol should be warm(ed) to above 125°F or set in the hot sun for several hours before mixing with gasoline. There are major advantages to this method. You know the exact proof of the alcohol being mixed, and can assure an exact percent blend. It also allows you to stockpile alcohol. There is also a safety factor involved.

Option Two. If you are certain of your still's ability to consistently turn out a given proof in a given run, you can run the end condensing coil right into the separation tank. The advantage is expediency—you have a one-step operation, and the alcohol is already warm. The major disadvantage is having a container of volatile gasoline so close to

Figure 10A. The Separation Tank

an operating still. Extreme caution is advised if you blend your gasohol this way.

Once the alcohol and gasoline are mixed and stirred, let the solution cool to any temperature below 90°F. This cooling process should take no longer than one hour. Open the valve and remove the water-alcohol that did not mix with the gasoline. Store this runoff fluid for redistillation during your next mash distillation run. Transfer the resultant gasohol to a storage tank or use right away. If stored, it must be in an approved, vented tank.

Gasohol Guideline.

1. Use 120 proof alcohol or higher for blending.
2. Use unleaded gasoline only. Remember that alcohol raises the octane rating one point higher for each 3.5 percent alcohol added, so you can use a cheaper, low-octane gasoline.
3. Warm fluids separate faster than cold fluids; but caution is advised so that the temperature not be too high.
4. The separation tank should be located away from your distilla-

tion apparatus. It should be removed from fire or sparks, and well ventilated.

5. Once mixed, let the gasohol blend sit for a few minutes to cool down. However, do not let it sit longer than an hour before removing the water-alcohol that did not mix with the gasoline, for osmosis will begin transferring alcohol out of the gasohol and into the water-alcohol solution.

11
How to Convert Your Car (or Truck) to Run on Total Alcohol

IT IS POSSIBLE to convert your car, truck, or tractor to run on total alcohol, or a very high percentage gasohol. The major decision you must make is whether to hire a qualified mechanic or to do it yourself. The former is by far the easier—and probably the safer—of your two options—take your car to a mechanic, leave it for a day or two, and pick it up ready to run on alcohol. However, his fee versus dollars saved by going alcohol must be considered. It will cost from $300 to $600 for a conversion, depending on the type of car you have, its engine or carburetor, and the mechanic. This means that you must save between $300 and $600 in fuel-cost difference within a reasonable period of time to offset your total investment. Conversely, to convert your car yourself requires motivation, knowledge, tools, and parts, but it will save you money—the major reason you are considering fuel alcohol anyway. The parts cost for conversion will be about $50 to $75, depending on your particular car and exactly what is needed for conversion. Your ability to scavenge junk-yard parts and to innovate and implement will have a bearing on your total do-it-yourself cost.

As stated in Chapter One: Making Your Own, pre-1975 vehicles are relatively easy to convert. Later models—1976 and up—are progressively more sophisticated, offering such features as fuel metering, on-board computer-controlled carburetors, altitude compensators, exhaust-gas oxygen monitors or sensors, and, of course, more elaborate and complicated pollution control devices and equipment. But don't let that scare you—yet.

It is not within the scope of this book to turn you into an ace mechanic. If you are not now the least bit mechanically inclined, and you have trouble changing the spark plugs and oil in your car, you should absolutely have a mechanic handle the conversion of your car for alcohol use. If you do the work yourself, you should have a mechanic check your work, because if the conversion is not done correctly, your engine may suffer irreparable damage. Your option, of course, is to conduct further research by reading books (such as the Peterson and Chilton books) that teach the rudiments of automobile mechanics—more specifically, the mechanics of your particular car/engine/carburetor. Consideration must be given to the time it will take you to learn, and to the tools you must buy to get you underway.

THE BASIC BASICS

There are three basic types of engines feasible for automobile and truck use: electric, external combustion, and internal combustion. To date, the internal combustion engine has been the most popular and widely used, for it has offered fewer problems than are intrinsic to either electric or external combustion types. An electric engine uses DC batteries for fuel. The problems experienced with electric-powered vehicles have been range (very limited) and the weight of the batteries (sometimes weighing as much as the balance of the car itself). An external combustion engine burns the fuel (wood, coal, oil, etc.) outside the engine, converts water to steam in a boiler, which drives the pistons and runs the vehicle—the same principle as the old locomotive steam engine. Past major problems with this type engine have been squeezing the bulky components into a regular-size car's engine compartment, and water regeneration—having to stop frequently to add the water lost when the steam exhausts itself. An internal combustion engine burns the fuel inside the engine. Fuel is fed into an engine cylinder, either by fuel injection or through a carburetor-manifold system, where it is compressed and ignited (exploded) to drive the piston and run the car. Of the three types, the internal combustion engine has always posed fewer problems and offered the greatest versatility. For at least the next decade, it will be the most practical engine for the majority of the populace.

Any internal combustion engine requires two things to run: fuel and fire (spark). If either is missing or malfunctioning, the car will not start, let alone run. Fuel delivery and spark delivery are two completely separate and different systems that must be working hand in hand for your vehicle to work at all. It is with the fuel system that you are most concerned.

How Your Car's Fuel System Works. Beginning with your fuel tank, usually made of galvanized sheet metal, fuel is delivered to the engine compartment through a fuel line, which is usually made of galvanized metal, copper, or stainless steel. Here, rubber(ized) or neoprene hoses connect to control the flow and direction (around corners and such) of the fuel to the fuel pump. From there, fuel is pumped into a float chamber built into the carburetor, where it waits and works on a demand basis. When the engine needs fuel, it draws it from the float chamber where it is forced through jets that shoot the fuel into the throat of the carburetor. There it vaporizes and mixes with air. This fuel-air mixture is sucked into the intake manifold, where it is distributed to the various cylinders as they demand fuel. The fuel-air mixture goes into the cylinder through an intake valve, where it is ignited, drives the piston down for power, and then leaves the cylinder via an exhaust valve. From there it goes into an exhaust manifold, where it then enters your exhaust system–catalytic converter or regular mufflers or a combination of the two.

There are, of course, variations on the basic fuel delivery system. Some cars have the fuel pump located within the fuel tank–nice and inaccessible. The location of fuel filters seems to vary at the whim of an engineer. Some filters are located in the fuel tank, some between the tank and the fuel pump, some between the pump and the carburetor, and some are built right into the carburetor. But these are minor concerns. The big concern is fuel vaporization–what happens to the fuel once it enters the carburetor.

Ignition. The major–if not only–function of the carburetor is to vaporize liquid fuel and mix it with air for subsequent combustion. Any carburetor will vaporize both gasoline and alcohol with relative ease. The major problem with alcohol is ignition. The flash point of gasoline goes as low as -50°F, which means if you cool gasoline to -50°F, you can light it with a match and it will burn without too much trouble. The flash point of ethanol is 70°F, a common home and summer temperature.

It is immediately apparent that as the temperature drops below 70°F, alcohol will be more difficult to ignite. This poses a very serious problem during the winter months when temperatures drop to zero and below, at least until the engine reaches operating temperature—130 to 160°F. Once the engine hits operating temperature, the flash point and autoignition temperature do not pose major problems for alcohol use. So how do you solve the problem?

Your major options are: preheat the alcohol, use a dual nonintegrated fuel system, use dual carburetion.

1. Preheating alcohol. This means raising the alcohol's temperature to above 70°F prior to feeding it into the carburetor, which can be done in a variety of ways. The major drawback to this solution is that when the warmed alcohol meets cold air in the carburetor and the vapor (air/fuel) hits the cold intake manifold, the ambient alcohol temperature will drop; and it is the temperature of the alcohol vapor that is critical for ignition. This does not mean you should not consider this possibility, but rather that it is not totally effective and is not a sure solution to cold-weather operation.
2. Dual nonintegrated fuel system. This uses both gasoline and alcohol. Nonintegrated means that they are not mixed at any time, from fuel storage to vaporization and ignition. In operation, you use gasoline to start the engine, and for warm-up. When the engine reaches operating temperature, you switch over to alcohol as the main running fuel. As long as the engine maintains operating temperature—or close to it—you can drive around town, park the car, and restart it on alcohol. However, when the engine temperature drops to below, say, 60 to 70°F, you must again use the gasoline for starting and warming up. An engine at idle speed (700 to 900 rpm) will consume about one gallon of gasoline per hour. If it takes five minutes for your car to reach operating temperature, you will then have twelve starts per gallon of gasoline. (This will vary depending on the size of your engine and the surrounding temperature.) If you only need to start your engine on gasoline twice a day, one gallon of gasoline will last about a week. A dual fuel system requires a reserve tank for gasoline storage; however, a one-gallon capacity should be sufficient. (See pages 146–148.)
3. Dual carburetion. With Option Two, one carburetor is used to receive either gasoline or alcohol. With dual carburetion, you have two separate carburetors—one to receive gasoline and one to receive alcohol. Dual carburetion works on the same principle as the dual fuel system—start the engine on gasoline and

switch to alcohol—but dual carburetion is far more efficient and effective. However, this option does require some mechanical expertise. Most engines have an intake manifold designed to receive one carburetor, with one barrel (one throat), or two, or four barrels. A four- or six-cylinder engine generally has a one- or two-barrel carburetor; an eight-cylinder engine usually has a two- or four-barrel carburetor. If your car has a two- or four-barrel carburetor, you cannot make it so that half of the carburetor runs on gasoline and the other half on alcohol. You must have two separate carburetors. Additionally, you will need a new (or used) intake manifold to accommodate the two carburetors, two fuel pumps, two hand-operated carburetor chokes, and assorted gaskets and linkage.

Another option is to use alcohol during summer months only, and use gasoline during winter months, since the major problems with fuel alcohol usage occur during cold weather. Problems generally start at 50°F to 60°F, and become more severe as the temperature drops. Depending on your region's median annual temperature, you could run your vehicle on alcohol for two thirds of the year with little or no problem, and when cold weather sets in, simply switch to gasoline.

HOW TO MAKE THE NECESSARY ADJUSTMENTS

Option One: Preheat the Alcohol. Any source of heat will suffice, although most have at least one disadvantage. Using the radiator, exhaust manifold, or the car's heating system lines all require engine heat before they will help heat the alcohol. However, an electric rear window defroster (RWD) provides almost instant heat and is viable. The RWD can be purchased in kit form at most auto supply shops and some chain discount stores at nominal cost. You must install a metal fuel feed line of at least 1 foot in length between the fuel pump and the carburetor float chamber. If the fuel pump is not in the engine compartment, install the metal line as close to the carburetor as possible without actually touching it. Use neoprene or rubber line for connections. Before you wrap the metal line with the RWD, coat the line with several thicknesses of a high-heat-resistant varnish. (The type used on electric motor armatures will do fine.) This coating prevents the grounding of the RWD to the line and subsequent short-circuiting of your electrical system. As you wrap the RWD on the line, coat the RWD with the same varnish, and be certain that no filaments (wires) touch each other. Once it is wrapped, apply several more coats of varnish for a final seal, and make the electrical connections according to the instructions that accompany the kit.

As a precaution, run the hot lead from the battery to the RWD through a fuse. Most cars have one or more blank terminals on the fuse box, which is usually located under the dashboard on the driver's side. If anything does go wrong, the fuse will kill the system before serious damage can occur.

Options. If you choose to use engine heat to assist in raising the temperature of the alcohol, the fuel line can be run through the top cap of the radiator. This involves removing the cap, drilling entry and exit holes, installing the line, and rebrazing the cap to the radiator body. This project is best tackled by a professional radiator (or welding) shop. In practice, the fuel line would run from the fuel pump into the radiator, out the other end of the radiator cap and into the float chamber. The line leading to the float chamber should be insulated to prevent loss of heat during its journey.

Exhaust manifold application simply means running a metal fuel line close to the exhaust manifold, without touching it, and into the float chamber. This is far less efficient than using the radiator, though far easier to do.

To use the car's heating system to raise the temperature of the alcohol, you must install a ½-inch copper tube directly into the inlet heating system water line. You then wire a metal fuel line to the copper jacket, and wrap and wire both with insulation to prevent heat loss and facilitate heat transfer from the jacket to the fuel line.

While none of the above three systems will help cold weather starts, they will assist in performance once the engine is running on alcohol.

At least one major U.S. auto manufacturer has a heating grid located between the carburetor and the intake manifold. As the fuel vapor passes through the grid, it is automatically preheated for ignition. Check into this feature, for it is an ideal set-up for fuel alcohol. If it is available for your carburetor brand and type, make the investment.

Option Two: Dual Nonintegrated Fuel System. The simplest method for utilizing a dual fuel system is to use a windshield washer spray set-up. This works the same as your regular windshield washer, except that instead of spraying water onto the windshield, it sprays gasoline into the carburetor. This can be purchased as a kit, or bootlegged from junk-yard parts. There are two types: one that is pressure-pumped using manual power, usually a small foot pump located inside the car near the headlight-dimmer foot switch; and an electric pump that operates at the flip of a switch.

Before you begin installation, decide on reservoir capacity. Most washer units have a one-quart capacity, which may or may not be enough for your particular needs. Adding a larger reservoir is easy provided you have the room in your engine compartment–some are extremely cramped. If there is no room, you can place the reservoir in the trunk and run a line forward. The pump and switch remain up front.

The pump is small enough so that it can be installed almost anywhere in the engine compartment. Place it close to the reservoir when possible. If the reservoir is in the trunk, place it on the same side of the car–left or right–when possible. Use sheet metal screws for attachment.

Install a two-position switch (on/off type) under the dashboard. Run a hot lead (positive or +) from the battery into the fuse box, and from there through the on/off switch. (The lead-in, or hot, wire attaches to one terminal, the lead-out wire leaves the switch through the other terminal.) Run the lead-out wire to the pump, and attach to one lead-in wire (red wire if they are colored black and red). Ground the other pump wire to any part of the car's body, and test the pump. If it does not squirt fluid, reverse the wires at the pump. NOTE: Some imports use red for the ground or negative wire color code; most all domestic autos use red for the hot or positive color code.

If the pump functions properly, measure a length of flexible tubing (1/32-inch inner diameter) from the pump to the carburetor so that there are no sags and no possibility of contact with the engine block. Use clips and run the tubing along the wheel well and/or firewall if necessary. Drill a hole slightly smaller than the tubing diameter through the carburetor throat wall. Be extra careful that you do not drill into any other part of the carburetor, for you could hit a fuel passageway and ruin the carburetor. This hole must be below the butterfly (a flat piece of metal in the carburetor's throat). If you place the hole above the butterfly, the fuel will have trouble getting into the intake manifold, for the butterfly is almost completely closed when the engine is cold. Force the tubing through the hole and into the throat at least ½ inch. This completes installation.

Testing, testing. Fill the reservoir with gasoline, disconnect the main fuel line into the float chamber and redirect it into a bucket (any fuel in your main system will pump when you run the engine), and start the engine to empty the float chamber. When the engine finally quits running, you are ready to test your new system.

Take a deep breath and flip the fuel switch to the on position. Engage the starter, and give the engine a few seconds to start. (Have faith. If the washer pump pumps, the engine will start.) If the engine runs for a few seconds, then stops or stalls, it is not getting enough fuel. Enlarge the tubing by 1/32 inch (.03125 inch). This means you must drill a larger hole into the throat. If, however, the engine races at high idle, the engine is getting too much fuel, and you must restrict the flow. Either go to a smaller-size tubing or crimp the tubing. Trial and error will give you the ideal match for your car. When you are satisfied with the starting and idling, attach the main fuel line, and get ready to run on alcohol.

Dual-Fuel Tips:

1. Do not use a regular electric gasoline fuel pump. This is a constant-flow pump, which is no good, because the flow rate per minute is too great.
2. Do not use a constant-flow washer pump.
3. Check the pump for operational condition before making any changes or hookups, particularly if it was junk-yard obtained.
4. Start with the smallest inner diameter fuel-feed tubing possible, and work your way up from there. Bear in mind that during operation, some alcohol will be metered into the carburetor during gasoline warm-up. There are no fixed standards for inner diameter of the tubing, pump pressure, and fuel flow rate, for cars are different according to the engine size, carburetion, and state of tune.

NOTE: Some fuel alcohol advocates recommend using a T-joint fitting fed by two reservoirs—one for alcohol, one for gasoline—to pump either alcohol or gasoline into the float chamber. For cold morning starts, you push a lever and feed gasoline into the float chamber, then, after warm-up, pull the lever to switch over to alcohol. It sounds logical, and feasible, but it will not work. When the alcohol and gasoline mix in the float chamber during switch-over, any water at all in the alcohol—and there is always a small amount—will immediately separate and settle out. Since the float chamber feeds the carburetor from the bottom, your engine will overdose on water. It will not run; it will not start.

Option Three: Dual Carburetion. This begins with a close look at feasibility. Some four-cylinder transverse engines (front-wheel drive) are so cramped for space in the engine compartment that there is barely enough room for the standard carburetor and intake manifold. If such is the case, forget dual carburetion.

Most rear-wheel drive cars have enough room in the engine compartment, regardless of the engine size, to receive another carburetor. Your task begins with parts acquisition. If you plan junk-yard acquisition, don't be surprised at the difficulty of finding an intake manifold set up to receive two separate carburetors. They are not that common, and you will probably have to buy this one item new from an auto parts outlet or car dealership. But before you do, give the junk yards a quick tour. You may get lucky.

Carburetor choice. If you have a one or two barrel carburetor, stay with the same. Since you already have one, you need only buy one more. However, if you have a four barrel carburetor, consider buying two used two barrel carburetors. During normal operation of a four barrel carburetor, only two of the barrels are functioning. The other two kick in only when the throttle lever reaches a certain point—usually three quarters throttle—or when you kick down to pass. Installation of the two barrel carburetors is easier, particularly the linkage; and the cost of the intake manifold will be less. If the carburetors have automatic chokes, they will have to be replaced with individual manual chokes—one for each carb.

Gaskets. Do not try to salvage any gaskets, either on your car or on any used parts you may buy. Gaskets are relatively inexpensive, and old gaskets have a tendency to leak somewhere—either letting air in or fuel out. Do not buy and/or use cork gaskets. Request an alcohol-resistant gasket when you buy. Vellomoid gasket material will suffice, when used with an alcohol-resistant gasket sealant. Do not use varnish for the gasket sealant; alcohol may eventually dissolve the varnish.

Installation. Once you have all necessary parts, installation is easy. Installing the intake manifold and carburetor units requires the simple removal of a few bolts and the old parts, and replacement with your new parts. Throttle linkage and manual chokes are a little more complicated.

Throttle linkage. There are two types of linkage: those that push, and those that pull. Some use a cable–like a choke cable–and some use solid linkage connected at pivot points. Regardless of the type, it may be necessary to lengthen or shorten the throttle rod or cable an inch or so because of the new position of the main carburetor–the one that will be running on alcohol. Plan on this prior to installation, for you may have to locate and buy a longer rod or cable. A hand-operated throttle cable for the gasoline carburetor will suffice, if, in fact, it is at all necessary.

Choke installation. The method and procedure for installation will vary depending on the type carburetor you have. In simplest form, disconnect the automatic choke, run the manual choke cable to the choke control plate (outside the carburetor), and connect it. Mount the push/pull knobs under your dashboard.

Fuel pump. Here you have two options. Use the existing pump on the engine for the main fuel (alcohol), and buy an electric fuel pump for the gasoline; or buy two electric pumps–one for each fuel. The advantage to the former is that you save the purchase price of a pump and the time for installation. The advantage to the latter is that you can kill the fuel to the alcohol carburetor with the flip of a switch. Even though you can control the air flow to the alcohol carburetor during warm-up with gasoline, some alcohol will find its way into the intake manifold. This may be more than your engine can handle–too rich of a fuel mixture–and may flood and stall the engine. The electric fuel pump installs in similar manner as the washer pump–two wires, two fuel lines (input and output), and two sheet metal screws.

Fine Tuning. When the mechanical installation is complete, it is time to set the engine up. It is assumed that the engine ran well before conversion–easy to start, no hesitation or sputtering, etc. You must now make adjustments on each carburetor individually, beginning with the gasoline carburetor. If the engine won't run on gasoline at this point, it will never run on alcohol. Eliminate the air and fuel flow to the alcohol carburetor. If necessary, place a piece of flat metal over the carburetor throat inlet.

When the car is running satisfactorily on gasoline, eliminate the fuel and air to that carburetor, and connect the alcohol carburetor. The major changes to be made to either carburetor will be to adjust the air/fuel mixture (ratio), and idle-screw setting.

For gasoline, the air/fuel ratio is 15 to 1 (15 parts air to 1 part fuel); and for alcohol it is 9 to 1, since alcohol contains so much more oxygen than gasoline. The gasoline carburetor should be pretty well set from its last use. However, it will be necessary to adjust the air screw on the alcohol carburetor to decrease the air/fuel ratio. Since alcohol has a lower heat value than gasoline, it will also be necessary to adjust the idle-screw setting, allowing more fuel in during idle rpm.

Most auto repair manuals give step-by-step procedures for idle screw and air/fuel ratio adjustments. These screws are easily accessible on the side of the carburetor. (NOTE: Some carburetors on late model cars have steel plug inserts to preclude adjustment. These must be drilled out to allow access to the screws. The legality of removing these plugs, or inserting them in the first place, is to be questioned.)

It may be necessary to advance engine timing one or two degrees, since alcohol burns less readily than gasoline. Timing advance retards the spark for a split second, allowing more complete compression (permitting the piston to reach absolute top dead center) before firing. It also allows the alcohol to absorb some engine heat just prior to ignition.

Advancing the timing requires a timing light, which can be borrowed from a neighbor or purchased for less than $20. The distributor, which distributes the ignition spark to the individual cylinders, is held in place with a single hold-down bolt. By loosening this bolt, you can rotate the distributor–clockwise or counterclockwise, depending on the car–and advance or retard the spark. Caution is advised here, for if ignition occurs too soon or too late, the engine will not run properly. If ignition is too soon, you will have incomplete burning of fuel, and, if the intake valves are partially open, backfiring through the intake manifold and/or carburetor. If it is too late, you will experience detonation; if an exhaust valve is partially open, it will backfire through the exhaust system, and may blow a hole in the muffler. It will also eventually burn the exhaust valves, necessitating replacement.

By using a timing light, you can set the engine to within a degree of factory specifications, and then work to either side of that setting one or two degrees with relative assurance that neither valve will be open during ignition.

It may be necessary to enlarge the carburetor fuel jets, but more likely it will not. This depends on the proof alcohol used to run your engine. Since the specific gravity (density) of pure alcohol (.79) is so close to that of gasoline (.70 to .79), alcohol will flow through the jets as readily as gasoline. However, as the alcohol proof drops with the addition of water, the density increases (water has a density of 1.00). The fuel jets must therefore be enlarged in direct proportion to the increase in fuel specific gravity.

As a rule of thumb, using 195 proof as a base unit, for every 10 percent decrease in alcohol proof (to a minimum proof of 120), increase (enlarge) the fuel jet size by 5 percent (to a maximum of 40 percent). Obviously this will vary according to your particular car, carburetion system, and engine compression.

The lower heat value of alcohol compared to gasoline has little or no bearing on fuel jet enlargement. Your engine works on a vacuum demand basis. As it needs more fuel, it draws as much as it needs until it is satisfied. The intrinsic balance to this is that alcohol will run your engine leaner, and the air/fuel ratio has been adjusted down.

Before you can enlarge the main carburetor fuel jets, you must remove the trappings surrounding the carburetor. These trappings include the air cleaner housing and related parts; air pollution hoses, both in and out; accelerator linkage; and choke accessories–either automatic-choke heating hoses or manual-choke cable. The carburetor itself can be held in one hand once the trappings are removed.

Most carburetors are built in two main pieces or shells, interlaced with holes, canals, screws, springs, diaphragms, plugs, and rods. With few exceptions there is an upper shell and a lower shell. Access to the main jets is usually as simple as removing the upper shell, exposing the fuel float chamber. The number of screws or bolts attaching the shells varies–there are usually six to eight–and they are very easy to remove and replace. The main jets will generally be located within the float chamber or in a recessed niche adjacent to the chamber. Physical identification is easy. The top view is of a round plug with a slot running the full width with a small hole in the center. (NOTE: Main jets on a Holley 4160 carburetor are located in a metering housing bolted to the side of the carburetor.)

Main jets can be removed with an ordinary screwdriver although a jet wrench is preferred. (NOTE: Ford's 1977 2700VV to 7200VV carburetors require special tools such as: jet wrench, jet plug remover and jet plug driver.) Remove the jets carefully and gently to avoid damage. Note which jet is left and which is right side, for they may vary in size or threading ever so slightly. Once enlarged, replace them in the same position–left or right.

Enlarging the jets. Although you can do this yourself, it is best to take the jets to a professional machine shop. The cost is exceptionally nominal and the end result far more desirable. If you choose to do it yourself or if there is no machine shop in your area, there are precautions to observe.

It is preferable to use a bench-type drill press with a hand-operated feed lever. In lieu of this, a small hand drill will suffice if you are careful. This requires a steady hand and a little patience.

Select the proper-size drill bit based on the rule of thumb alcohol-water percentage factor described on page 152, and carefully enlarge each jet to specification. Use extra fine emery cloth (preferably wet) to dress the face of the plug in order to remove any burrs or scars from boring.

To immobilize the plug while boring, it is best to use two pieces of soft wood (pine will do) at least ¼ inch thick, placed on opposite sides of the plug (one on each side). Gently but firmly clamp this with a bench vise or C clamp so that it is held securely but without damaging the threads of the plug.

While boring, exercise extreme caution so that nothing moves in other than in a straight line. If the piece or drill wobbles, it will create an oblong and oversize jet hole, which will not function properly when reinstalled. The hole must be absolutely straight and true throughout.

DO NOT: damage the jet plug threads, bore the hole too large, or bore the hole at an angle. DO NOT: use wood drill bits, or hold the plug in your bare hand while boring.

When boring is complete, reinstall the plugs into their respective seats, and reassemble the carburetor using a new gasket and alcohol-resistant gasket sealant.

FUEL INJECTION

Fuel injection conversion is best left to the experts, unless you are very familiar with the system. Although there are two major types—gasoline and diesel fuel—both function on about the same principles.

A fuel injection engine is a high-compression engine using a heavy-duty block and internal components. Fuel delivery is usually accomplished by a fuel pump that pumps the fuel to an injection pump. Most have a second booster fuel pump in the fuel tank (or just outside of it) to prevent vapor lock on a fuel return system. From the injection pump, fuel is delivered to cylinders via fuel lines commonly called rails. The fuel is then injected into the cylinders (or intake manifold)

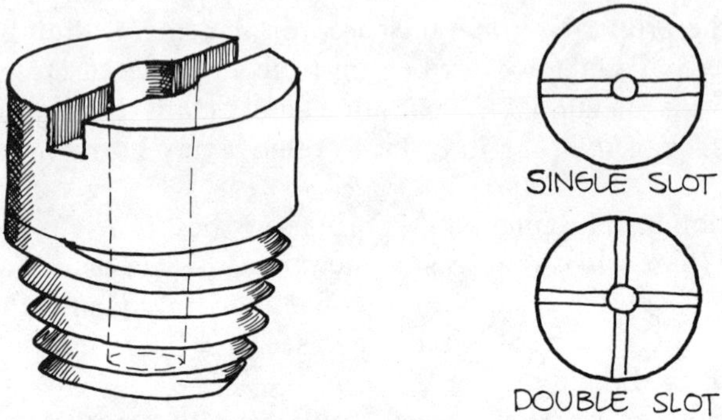

Figure 11A. The Carburetor Fuel Jet Plug

through metering valves or injectors. Ignition occurs during the compression stroke of a particular cylinder by the heat generated by the compression of fuel and air. Temperatures may reach as high as 1000°F. Glow plugs are used to start the engine and run it until heat-generated compression takes over. There are two basic types of glow plugs: controlled pulsing current, and constant current. Strategically located sensors control the main functions of fuel/air ratio and ignition characteristics.

Before attempting a fuel-injection alcohol conversion, it is most advisable to consult a mechanic and do extensive book research on your particular engine.

In lieu of conversion, alcohol is an ideal fuel extender for fuel injection systems. You would make diesehol in the same manner as gasohol–using a separation tank and in the same basic proportions–although you can use a higher percentage of alcohol because of the high heat generated in a diesel system.

CAUTION: The diesel fuel injection pump is the heart of a diesel engine, and unfortunately most are extremely sensitive to water. Before you even think of running a diesel engine on fuel alcohol, check with the manufacturer of the vehicle or a topnotch mechanic, since fuel alcohol will always contain a certain amount of water.

12
Using What You Have Made to Fuel Your Home Heating System

ALTHOUGH YOU might not want to use alcohol to run your car, or you might not want to bother making gasohol, you almost certainly can use fuel alcohol to heat your home—and you don't have to mess with blending alcohol with gasoline, because many home heating systems can run on alcohol.

Heating systems, regardless of the type, seem complex simply because of their size. Some older systems (octopus) consume the space of a subcompact car. There are, of course, the smaller units, such as those found in mobile homes. The heart of most heating systems can really be held in two hands, and most heating systems are not that complicated or difficult to work on. In fact, very simple household tools can be used—screwdriver, pliers, wrenches.

Heating systems have the following characteristics, some of which are similar, if not identical, in nature, scope, and function:

1. A main or central thermostat located somewhere in the house (building) controls the ambient temperature of the dwelling by turning the furnace on and off. Some electric heating systems have zone temperature control—a thermostat in each room for individual heat control.
2. Source of fuel: either on-site stored (propane gas, coal, wood, or oil), or delivered (natural gas or electricity).
3. Fuel delivery: gravity- or pump-fed (oil); screw- or hand-fed (coal or wood); pressure-fed (natural gas and propane gas); or land-line-fed (electricity).
4. Fuel regulator: gas uses a pressure regulator and metering

valve; oil uses a low-to-high-pressure pump, sometimes of positive displacement type; electricity uses a rheostat to control the flow of electricity to the heating coils; and wood or coal are generally personally regulated or tended. (The exception would be a screw-fed coal furnace. When the temperature drops to a preset level, the screw is automatically turned on and feeds more coal into the furnace.)

5. Fuel ignition: gas uses a pilot light, which always burns; oil uses one of two types of ignitors, a constant fire or an intermittent fire ignitor; coal and wood are usually fired by hand (just like being in the scouts); and electricity simply flows when switched on.
6. With the exception of wood and coal, most furnaces have automatic controls that regulate turning on and off, and safety controls that shut the system down whenever problems develop. Those furnaces equipped with automatic controls usually have a manual override control that allows you to manually start or stop the furnace.
7. Heat distribution. Regardless of the heat source, there are three major methods of heat distribution: hot air, hot water, and steam. In a hot air system, the air is heated and rises through ductwork for distribution throughout the dwelling. The two types of hot air systems are gravity and forced air. Gravity works on the principle that hot air rises and cold air falls. As the hot air heats the house, it cools, and returns to the furnace to be heated again. Forced air uses a fan or blower to force the hot air up into the dwelling, drawing fresh air either from the basement or a cold-air return.

   Hot water and steam heating systems use pipes and plumbing to distribute the furnace heat to radiators—either bulky stand-up types or baseboard types. With hot water heat, the water is heated in a boiler, and then pumped throughout the system with a recirculating pump. As it gives up its heat, the cold water is returned to the furnace. Steam heat works strictly on pressure and condensation. As the steam is generated, it forces its way throughout the system to the condensing units (radiators). As the steam gives up its energy, it converts back to the heavier liquid form, which returns to the furnace by gravity, where it is heated to steam again.

If you have a fuel-oil-fired hot-water or steam heating system, you will have very little trouble switching over to an alcohol-fired heating system. Your main considerations are the fuel jets (spray nozzles) and perhaps the ignitor. The fuel jets may need to be changed to the next larger size, which is no major operation. If you have a constant-fire ignitor, no change is necessary. If, however, you have an intermittent ignitor, it must be changed to a constant-fire ignitor. Regardless, your conversion to fuel-alcohol heating is relatively simple and inexpensive.

If you have a hot water heating system fired by any other fuel, you must consider the cost of ripping out what is there and replacing it with the equipment necessary to spray and ignite alcohol; not impractical, but a little more expensive than the changes required for conversion of a fuel-oil-fired hot-water heating system. Most of what you need—boiler, expansion tank, and plumbing—is already there. Consideration depends on the cost to modify your particular unit and/or whether or not you do it yourself. It is worth getting a professional estimate on total cost.

If you have any other type heating system, the cost of installing a new system start to finish may preclude using alcohol to heat your home. It may be twenty years before you break even when you consider the cost of installation against the savings of making and using your own fuel. Of course, at any time your existing system malfunctions or needs replacement, fuel alcohol may become feasible for you. You might as well replace your existing system with a system that permits you to make your own fuel.

HOW TO CONVERT YOUR PRESENT SYSTEM

First and foremost, all components of your heating system must be operational. If something won't work on fuel oil, it surely will not work on fuel alcohol. Repair or replace all defective parts before you begin. Next, purge the system of fuel oil. Most ideally, you will purge all fuel oil from the system by using the oil for heating purposes. Why waste fuel that can be used? The entire system must be purged, from the fuel tank (reservoir) right down to the pump. A few drops of fuel oil here and there will not really hurt; the alcohol will eventually clean the system out top to bottom.

Depending on your particular heating unit, it may or may not be necessary to change the spray jet nozzles. Before you waste the time, effort, and money to install new nozzles, try to run your heating unit on alcohol. If, for whatever reason, you experience any problems with fuel delivery and/or ignition, it will then be necessary to change the nozzles to the next larger size. Here's how it's done.

Remove the tray assembly, which holds the pump and spray jet nozzles. With either pliers or wrench remove the nozzle(s). The nozzle consists of a brass plug with a screen (filter) preceding it. The plug has a hole bored through it for fuel passage, and is concave on the front face so that as fuel shoots out, it is forced into an umbrella-shaped fuel mist. While some advise simply boring a larger hole in the plug, this is not advisable, for it will scar the plug's face and inhibit if not ruin the umbrella misting effect. If the fuel is not misted, and squirts out in a confined stream, ignition is literally impossible. Replace the plug with a new one, the next larger size. (Sizes will vary depending on the particular system, so take the old one to a plumbing or hardware store for comparison.)

When the new plug is installed, it is time to test for function and misting. This should be done before the tray assembly is reinstalled in

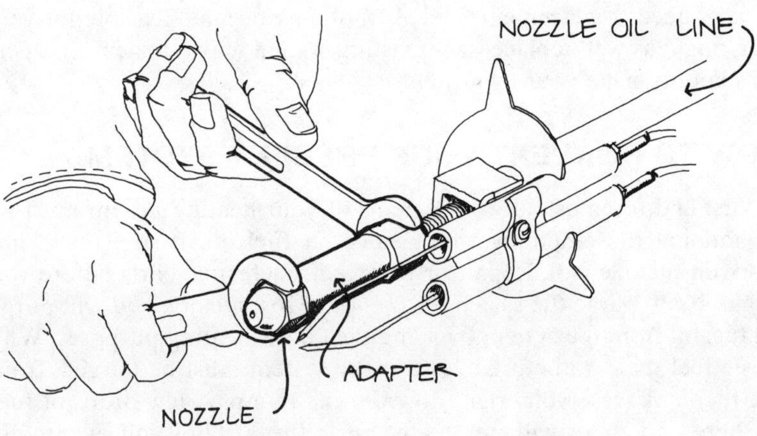

Figure 12A. The Tray Assembly

the furnace. Rather than load the fuel tank with alcohol, disconnect the fuel feed line from the pump and attach a neoprene or rubber hose. Place the open end in a container of alcohol and turn the pump on. It will take a few seconds for the alcohol to feed into the pump.

CAUTION. Testing this way should be done only if the area is well ventilated, and only if there is absolutely no possibility of alcohol ignition—no sparks or open flame present. This includes the ignitor within the furnace. Unless you disconnect the ignitor, when you turn on the pump you will also turn on the ignitor.

If the pump and nozzle mist the alcohol to your satisfaction, this conversion phase is complete. If it squirts, but does not mist, use a small piece of emery cloth (sandpaper) to chamfer the edge of the plug's outlet hole. If it still will not mist, try the old plug.

Ignition. Most oil-fired furnaces have an intermittent ignitor, primarily for energy conservation. Once the oil ignites, the heat and flame inside the furnace continue ignition of fresh fuel as introduced. Any fuel that does not immediately ignite drops to the bottom of the tank (furnace) and is eventually burned from the sheer heat of the furnace. Although an intermittent ignitor may work with fuel alcohol, a constant-fire ignitor is more advisable for two reasons. Depending on the ambient temperature and the amount of water in the alcohol, your fuel alcohol will be slightly more difficult to ignite than fuel oil. Also, a constant ignitor diminishes the chances of a flameout.

Most oil furnaces have a heat sensor in the vent duct. When the pump kicks on, the sensor allows five to ten seconds to feel (sense) heat. If it does not, it will shut the furnace (pump) down. If, while using fuel alcohol, the flame goes out, the heat within the furnace will fool the vent heat sensor, and the pump will continue to pump raw alcohol into the furnace. For this reason alone, a constant ignitor is most desirable. If you have an intermittent ignitor, change it to a constant-fire type.

Most oil furnaces do not have a drain in the bottom of the firebox, since any oil that may accumulate on the bottom will eventually burn out. However, with fuel alcohol, water is inevitably going to be present. While most of it will vaporize to steam and be vented out of the system, some of it may collect on the bottom of the firebox and, if allowed to sit, eventually cause rusting. This depends to a great extent on the proof alcohol you are using. Higher proofs will be less susceptible to this potential problem. If you plan to use less than 160 proof alcohol for fuel, drill a drain hole in the bottom of the firebox and attach a heat-resistant tube to dispose of any water accumulation.

If all checks out to this point, reassemble the tray assembly, reconnect the fuel line, and fill your tank with alcohol for an operational check. Turn the system on and allow enough time for air to clear the feed line and pump. NOTE: The safety vent sensor may shut you down when it does not sense heat. Until you are operational, you can either bypass or disconnect the sensor. If the system does not work, it is time to backtrack. Ask yourself:

1. Is fuel getting to the pump?
2. Is the pump spraying fuel?
3. Is the ignitor working?
4. Is the vent sensor shutting you down?
5. Most important, is the alcohol of high enough proof to burn? Less than 100 proof just will not work that well.

Options. In lieu of converting to run on 100 percent alcohol, you may choose to use your alcohol as a fuel extender. Unlike making gasohol, and unlike running your car on alcohol, any percent blend will work. Fuel phase separation must still be handled. A little water settling out will not hurt. It will simply be pumped into the firebox and, when ignition occurs, be converted to steam and vented out of the system. However, if you use a large quantity of low-proof alcohol, the water may stream long enough for the vent sensor to shut you down. If this is the case, it is best to install a water drain on the extreme bottom of your fuel storage tank. Whenever you mix, drain.

If you have a water or steam system fired by a gas or solid fuel, conversion is feasible, and may not cost a bundle, for you need not install brand-new liquid-fuel equipment. Many people, disenchanted and/or disgusted with ever-increasing oil prices, are switching to other types of fuel and furnaces–coal, gas, or electric. Their existing systems may be in excellent condition, but they want to get out from under the oil barrel. So watch the newspaper For Sale ads, contact heating and plumbing contractors, and check junk yards frequently. A used heating unit is just about as good as brand-new, and the cost will run anywhere from 10 to 40 percent the price of a new unit. If you have no immediate success finding a used unit, consider placing an advertisement in your local newspaper want ads stating exactly what you want.

Bear in mind that you will definitely need help hauling anything you may find, and that installation is not a one-man job. If you are the least bit skeptical about doing it yourself, hire a professional.

Another option is to convert portable, non-electric space heaters to fuel-alcohol use. The cost of these portable units is nominal compared to the impact they could have on your monthly heating bill, particularly since you can make your own fuel. This is not a total answer to your heating needs, but more a supplement; and every little bit helps.

The only conversion necessary to utilize these portable heaters is to switch from one fuel (kerosene) to another (alcohol), since they use a wick for fuel delivery and burning.

Conversions and Factors

1 pound = 453.6 grams
1 ton = 2000 pounds
1 gallon ethanol = 6.6 pounds
1 gallon water = 8.33 pounds
1 liquid gallon = 4 quarts = 231 cubic inches
8 gallons (dry measure) = 1 bushel
1 bushel = 56 pounds
1 bushel displaces 6.656 gallons of water
1 cwt = 100 pounds (hundredweight)
1 cubic foot = 7.48 liquid gallons
1 cubic foot = 62.36 pounds of water (at 60°F)
1 cubic foot = 1728 cubic inches
1 acre = 43,560 square feet
1 Btu (British thermal unit) = 252 calories
1 Btu = heat required to raise 1 pound of water 1 degree Fahrenheit
To convert from °F to °C, subtract 32 and then multiply by five ninths.
To convert from °C to °F, multiply by nine fifths and then add 32.
To compute square footage, multiply length times width.
To compute cubic footage, multiply length times width times height.
Per-bushel-yield, gallons of ethanol = % carbohydrates in bushel times 6.656 times .50
Per-acre-yield, gallons of ethanol = gallons per bushel times bushels per acre

Cost per gallons of ethanol = cost per bushel divided by gallons per bushel
Freezing points: ethanol = -173°F, water = 32°F, gasoline = -70°F
Boiling Points: ethanol = 173°F, water = 212°F, gasoline = 80°F to 440°F

Thermal Properties

Btu per pound: ethanol = 12,800, gasoline = 20,260, fuel oil = 20,000
Btu per gallon: ethanol = 84,400, gasoline = 124,800
Autoignition temperature: ethanol = 685°F, gasoline = 430 to 500°F
Flash point: ethanol = 70°F, gasoline = –50°F
Octane Ratings at pump: ethanol = 98 to 100, gasoline = 86 to 90+

Contact Information

YEAST AND ENZYME MANUFACTURERS AND DISTRIBUTORS

Anheuser-Busch, Inc.
721 Pestalozzi Street
St. Louis, MO 63118
314-577-2000

Biocon, Inc.
216 Midland Avenue
Lexington, KY 40507
606-254-0517

Chemapec, Inc.
230 Crossways Park Drive
Woodbury, NY 11797
516-364-2100

Enzyme Development Corp.
210 Plaza
New York, NY 10001
212-736-1580

Fermco Biochemicals, Inc.
2638 Delton Lane
Elk Grove Village, IL 60007
312-595-3131

G. B. Fermentation Inc., Inc.
One North Broadway
Des Plaines, IL 60061
312-827-9700

Grove Engineering
1714 Gervais Avenue
North St. Paul, MN 55109
612-777-8545

* Diamond V. Mills, Inc.
436 G. Avenue, NW
P. O. Box 4408
Cedar Rapids, IA 52407
319-366-0745

Norbert Haverkamp
Compost Making Enzymes
Rural Rte. 1, Box 114
Horton, KS 66439
913-486-3302

Miles Laboratories, Inc.
Enzyme Products Div.
P.O. Box 932
Elkhart, IN 56515
219-564-8111

Nova Laboratory, Inc.
59 Danbury Road
Wilton, CT 06897
203-762-2401

Rohm & Haas
Independence Mall West
Philadelphia, PA 19105
215-592-2517

Schwarz Service International, Ltd.
230 Washington Street
Mt. Vernon, NY 10551
914-664-1100

*Tall Grass Research Center
Rte. 2, Box 21
Formosa, KS 66942
913-794-2236

* Universal Foods Corporation
Fermentation Products Div.
433 E. Michigan Avenue
Milwaukee, WI 53201
414-271-6755

* Best sources for small-scale operations.

STATE-BY-STATE GUIDE TO FUEL ALCOHOL INFORMATION

ALABAMA: Energy Management Board, 3734 Atlanta Hwy., Montgomery, AL 36130

ALASKA: Community Information Center, P.O. Box 1267, Fairbanks, AK 99707

ARIZONA: Office of Economic Planning and Development, P.O. Box 126, Williams, AZ 86046

ARKANSAS: Department of Energy, 3000 Kavanaugh St., Little Rock, AR 72205

CALIFORNIA: Energy Commission, 1111 Howe Ave., MS 68, Sacramento, CA 95825

COLORADO: Department of Agriculture, 1525 Sherman St., 4th Fl., Denver, CO 80203

CONNECTICUT: Office of Policy Management and Energy Division., 80 Washington St., Hartford, CT 06115

DELAWARE: Energy Office, P.O. Box 1401, 114 Water St., Dover, DE 19901

FLORIDA: Governor's Energy Office, 301 Bryant Bldg., Tallahassee, FL 32301

Department of Agriculture, Mayo Bldg., Tallahassee, FL 32301 (Alternate)

GEORGIA: Office of Energy Resources, 270 Washington St., SW, Rm. 615, Atlanta, GA 30334

HAWAII: Department of Planning and Economic Development, P.O. Box 2359, Honolulu, HI 96804

IDAHO: Office of Energy, State House, Boise, ID 83702

ILLINOIS: Illinois Liquor Control Commission, 160 North LaSalle St., Chicago, IL 60601

INDIANA: Department of Commerce, 440 North Meridian, Indianapolis, IN 46204

IOWA: Iowa Development Commission, 250 Jewett Bldg., Des Moines, IA 50309

KANSAS: Energy Office, 214 West 6th St., Topeka, KS 66603

KENTUCKY: Department of Energy, Box 1188, Ironworks Pike, Lexington, KY 40578

LOUISIANA: Department of Natural Resources, P.O. Box 44156, Baton Rouge, LA 70804

MAINE: Office of Energy Resources, 55 Capitol St., Augusta, ME 04330

MARYLAND: Energy Policy Office, 301 West Preston St., Suite 1302, Baltimore, MD 21201

MASSACHUSETTS: Executive Office of Energy Resources, 73 Tremont St., Boston, MA 02108

MICHIGAN: Energy Administration, Department of Commerce, P.O. Box 30228, Lansing, MI 48909

MINNESOTA: Energy Agency, 150 East Kellogg Blvd., St. Paul, MN 55101

MISSISSIPPI: Office of Energy, Suite 228, Barefield Complex, 455 North Lamar, Jackson, MS 39201

MISSOURI: Department of Natural Resources, Energy Division, P.O. Box 176, Jefferson City, MO 65102

MONTANA: Department of Natural Resources, 32 South Ewing, Helena, MT 59601

NEBRASKA: Agricultural Products Industrial Utilization, 3rd Fl., 301 Centennial Mall, South, Lincoln, NE 68509

NEVADA: Department of Energy, 1050 East Will, Suite 405, Carson City, NV 89710

NEW HAMPSHIRE: Governor's Council on Energy, 2½ Beacon St., Concord, NH 03301

NEW JERSEY: Department of Energy, 101 Commerce St., Newark, NJ 07102

NEW MEXICO: Energy and Minerals Department, Box 2770, 113 Washington Ave., Santa Fe, NM 87501

NEW YORK: Bureau of Resource Development Energy Office, Agency Bldg. #2, Rockefeller Plaza, Albany, NY 12223

NORTH CAROLINA: Energy Division, Department of Commerce, P.O. Box 25249, Raleigh, NC 27611
Department of Agriculture, 1 West Edenton St., Raleigh, NC 27611 (Alternate)

NORTH DAKOTA: Energy Management and Conservation, 1533 North 12th St., Bismark, ND 58501

OHIO: Department of Energy, 30 East Broad St., 34th Fl., Columbua, OH 43215

OKLAHOMA: Department of Energy, 4400 North Lincoln Blvd., Suite 251, Oklahoma City, OK 73105

OREGON: Department of Agriculture, 636 Capitol St., Salem, OR 97310

PENNSYLVANIA: Department of Agriculture, 201 Agriculture Bldg., 2301 North Cameron St., Harrisburg, PA 17120

RHODE ISLAND: Cooperative Extension Services, Woodward Hall, University of Rhode Island, Kingston, RI 02881

SOUTH CAROLINA: Governor's Office of Energy Resources, 1122 Lady St., Columbia, SC 29201

SOUTH DAKOTA: Office of Energy Policy, Capitol Lake Plaza, Pierre, SD 57501

TENNESSEE: Tennessee Energy Authority, Capitol Blvd. Bldg., Suite 707, Nashville, TN 37219

TEXAS: Texas Energy and Natural Resources Advisory Council, 411 West 13th St., Austin, TX 78701

UTAH: Energy Office, 231 East 400 South, Suite 101, Salt Lake City, UT 84111

VERMONT: Energy Office, State Office Bldg., Montpelier, VT 05602

VIRGINIA: Department of Agriculture and Consumer Services, 203 North Governor St., Richmond, VA 23219

WASHINGTON: Energy Office, 400 East Union St., Olympia, WA 98504

WEST VIRGINIA: Fuel and Energy Office, 1262½ Greenbriar, Charleston, WV 25311

WISCONSIN: Division of State Energy, 1 West Wilson, Rm. 201, Madison, WI 53702

WYOMING: Governor's Energy Office, Capitol Hill Office Bldg., Cheyenne, WY 82002

ADDITIONAL FUEL ALCOHOL INFORMATION SOURCES

National Alcohol Fuels
Information Center
1617 Cole Blvd.
Golden, CO 80401
800-525-5555

National Gasohol Commission
521 South 14th Street, Suite 5
Lincoln, NE 68508
402-475-8044

Technical Information Center
U.S. Department of Energy
P.O. Box 62
Oak Ridge, TN 37830

Mr. Scott Sklar
National Center for
Appropriate Technology
1522 K St. NW
Washington, DC 20005

Iowa Corn Promotion Board
402 West Tower
1200 35th Street
West Des Moines, IA 50265

SOURCE CONTACTS FOR BUYING A READY-MADE OPERATION

ACR Process Corporation
602 Green Street
Champaign, Illinois 68120

Agri-Stills of America
3550 Great Northern Avenue
Route 4
Springfield, Illinois 62707

Agrodyne, Inc.
P.O. Box 934
Idaho Falls, Idaho 83401

Alcodynamics Corporation
P.O. Box A
Carlisle, Iowa 50047

Alcohol Plant Supply Co.
P.O. Box 248
Sherwood, Oregon 97140

Alcohol Technology, Inc.
P.O. Box 1489
Rockdale, Texas 76567

Arlon Industries, Inc.
P.O. Box 347
Sheldon, Iowa 51201

Brannon, Camp & Wimbish, Inc.
6755 Peachtree Industrial Blvd.
Suite 108
Atlanta, Georgia 30360

Double A Products
Division of Ad-Art, Inc.
P.O. Box 1107
Albert Lea, Minnesota 56007

Double L Manufacturing, Inc.
P.O. Box 533
American Falls, Idaho 83211

Easy Engineering
3351 Larimer Street
Denver, Colorado 80205

Energy Independence Corporation
Montrose, Minnesota 56308

Ethanol International, Inc.
1372 South Filmore
Denver, Colorado 80210

Expert Industries, Inc.
80 19th Street
Brooklyn, New York 11232

McEver Engineering, Inc.
6363 Richmond Avenue
Houston, Texas 77057

Middle State Manufacturing Co.
Box 788
Columbus, Nebraska 68601

Moier Construction
511 North Cedar
P.O. Box 381
Owatonna, Minnesota 55060

Moore Engineering Co., Inc.
1801 Falls Avenue
Waterloo, Iowa 50701

Rochelle Energy Development, Inc.
601 Sixth Avenue
Rochelle, Illinois 61068

Schmitt Energy Systems
Route 2
Hawkeye, Iowa 52147

Silver Engineering Works, Inc.
3309 Blake Street
Denver, Colorado 80203

Solar Fuel Company
Box 423
Dubuque, Iowa 52001

United International
P.O. Box 11
Buena Vista, Georgia 31803

United International, Inc.
United International of Buena Vista
P.O. Box 271
Buena Vista, Georgia 31803

Index